Industrial Stu
for Building Stu

Industrial Studies for Building Students

SECOND EDITION

RAYMOND W. STACEY
MCIOB

Swindon College

OXFORD

BLACKWELL SCIENTIFIC PUBLICATIONS

LONDON EDINBURGH BOSTON

MELBOURNE PARIS BERLIN VIENNA

Blackwell Scientific Publications
Editorial Offices:
Osney Mead, Oxford OX2 0EL
25 John Street, London WC1N 2BL
23 Ainslie Place, Edinburgh EH3 6AJ
3 Cambridge Center, Cambridge, Massachusetts 02142, USA
54 University Street, Carlton Victoria 3053, Australia

Other Editorial Offices:
Librairie Arnette SA
2, rue Casimir-Delavigne
75006 Paris
France

Blackwell Wissenschafts-Verlag
Meinekestrasse 4
D-1000 Berlin 15
Germany

Blackwell MZV
Feldgasse 13
A-1238 Wien
Austria

First Edition published under the title *Industrial Studies for Building Craft Students* by Collins Professional and Technical Books 1985
Second Edition published by Blackwell Scientific Publications 1992

Typeset by CG Graphics, Aylesbury
Printed and bound in Great Britain by Hartnolls, Bodmin, Cornwall

DISTRIBUTORS

Marston Book Services Ltd
PO Box 87
Oxford OX2 0DT
(*Orders:* Tel: 0865 791155
Fax: 0865 791927
Telex: 837515)

USA
Blackwell Scientific Publications, Inc.
3 Cambridge Center
Cambridge, MA 02142
(*Orders:* Tel: 800 759-6102
617 225-0401)

Canada
Oxford University Press
70 Wynford Drive
Don Mills
Ontario M3C 1J9
(*Orders:* Tel: 416 441-2941)

Australia
Blackwell Scientific Publications (Australia) Pty Ltd
54 University Street
Carlton, Victoria 3053
(*Orders:* Tel: 03 347-0300)

British Library
Cataloguing in Publication Data

Stacey, Raymond W.
Industrial studies for building students.
2nd ed.
I. Title
690

ISBN 0-632-02906-4

Library of Congress
Cataloging-in-Publication Data

Stacey, Raymond W.
Industrial studies for building students / Raymond W. Stacey.— 2nd ed.
p. cm.
Rev. ed. of: Industrial studies for building craft students. 1st ed. 1991.
Includes index.
ISBN 0–632–02906–4
1. Building. 2. Construction industry— Great Britain.
I. Stacey, Raymond W. Industrial studies for building students. II. Title.
TH146.S75 1992
690′0941—dc20

91–42572
CIP

This book is dedicated to my grandparents
Joseph and Martha Stacey

Contents

Preface

The aim of this text book is to provide a broad based introduction to some of the diverse aspects of the construction industry. It has been written to satisfy the requirements of those studying the Industrial Studies component common to all City and Guilds of London Institute Construction Craft Courses. It will also provide a broad based introduction to the construction industry and building crafts for those taking construction NVQs (National Vocational Qualifications). However, it will also be eminently suitable for those students on other courses with a construction content such as Certificates of Pre-Vocational Education (CPVE), Technical and Vocational Educational Initiative schemes (TVEI), General Certificate in Secondary Education (GCSE) as well as the Business and Technician Education Council (BTEC) First Certificate and Diploma Courses in Construction.

I would like to thank my friends and colleagues, both in teaching and the construction industry, for the support, advice and encouragement given during the writing of this book. Also, I must express my gratitude for the endorsement of those who have used this book and found it to be an invaluable aid to their studies, teaching and other enterprises.

Raymond W. Stacey

1 Modern Buildings

Modern societies need many different types of building. Each building has to be specially designed to provide the facilities needed for its use or purpose. The uses of buildings can be broadly classified into six main areas:

Residential.
Community.
Social.
Industrial.
Commercial.
Civic.

RESIDENTIAL BUILDINGS

One of primitive man's first activities must have been the provision of shelter for his family. He needed this shelter to protect his family from the bad weather, wild animals, his enemies, and to provide storage for his food and possessions. The need to provide shelter and security has been one of man's major occupations down the centuries to the present day. It is still a very important activity today.

The size, type, quality, quantity and design of accommodation that man has been able to provide for his family, animals and possessions have always been governed by the climate, the standard of society, the state of the economy, the types of material available, and the state of technology at the time of construction.

The individual used to be the driving force behind the provision of domestic dwellings. Now the state in the form of the local authority and speculative developers is the main builder of the dwellings in which we live today. The individual's only concerns are matching his personal needs and his financial resources with the quality, type and price of the accommodation available. If the purchase of a dwelling is beyond the financial resources of the individual, he has to resort to renting accommodation from the local authority or a private landlord.

The construction of domestic dwellings is consequently for one of the following groups of purchaser:

Owner occupiers.
Local authorities.
Private landlords.

TYPES OF DWELLING

Dwelling: A dwelling is a place of residence, a home where people live.

A wide range of dwellings are needed to satisfy the demands of society. For example: we need small one-bedroom dwellings for single people; dwellings with easy access from ground level for the elderly and the handicapped; dwellings for the standard-sized family; and dwellings with many bedroom for large families.

Terms used to describe the different types of dwelling:

Bungalow: A bungalow is a single storey dwelling. The name comes from India and is used to describe 'a dwelling of one storey'.

Chalet: A chalet is a bungalow that has a room, or rooms, built into the roof space. Purpose-built chalet bungalows have steeply pitched roofs. The steep pitch enables the greatest amount of floor area to be obtained in the roof space.

House: A house is a dwelling of more than one storey, i.e. it has more than one floor level.

Town house: A town house is a dwelling that is detached, or terraced, and has three or more floors. Town houses are built on very small plots of land in built-up areas. The high cost of land dictates that the area on which each dwelling is built must be as small as possible. Consequently, the dwelling has to have extra floors built on top to obtain the required floor area needed by the occupants.

Flat: A flat is a suite of rooms on one floor forming a self-contained dwelling in a multi-storey block of similar dwellings.

Maisonette: A maisonette is a suite of rooms forming a dwelling which occupies more than one floor in a multi-storey block of dwellings.

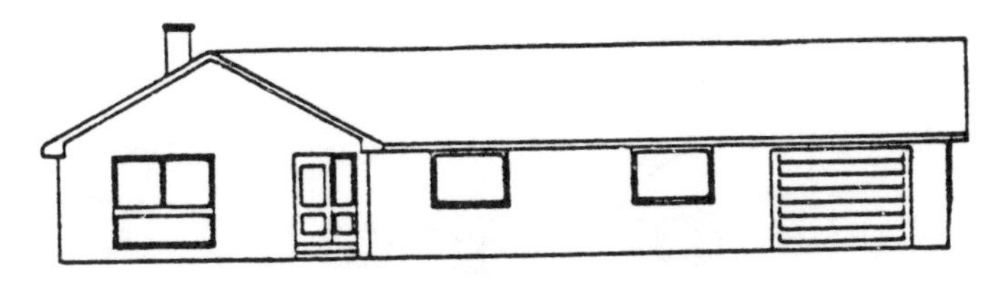

Detached

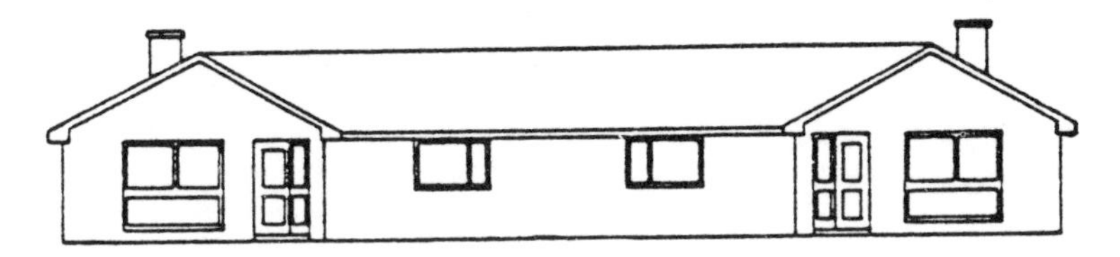

Semi-detached

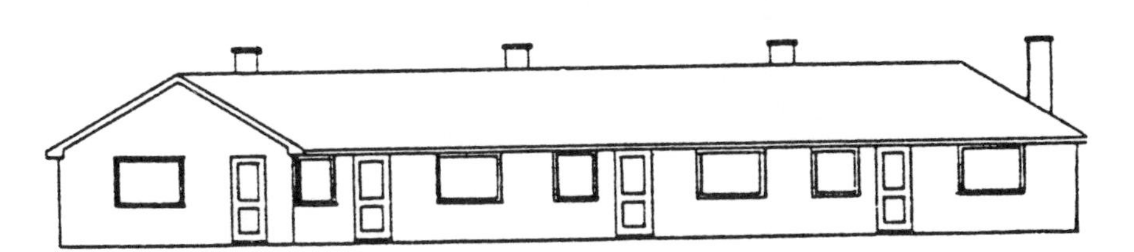

Terraced

Chalet

Bungalows.

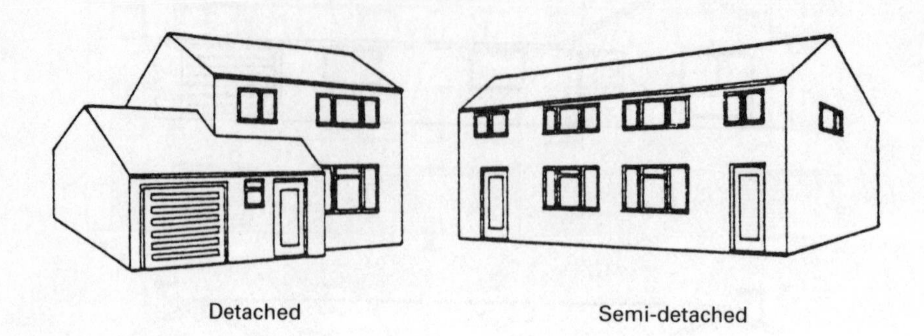

Detached Semi-detached

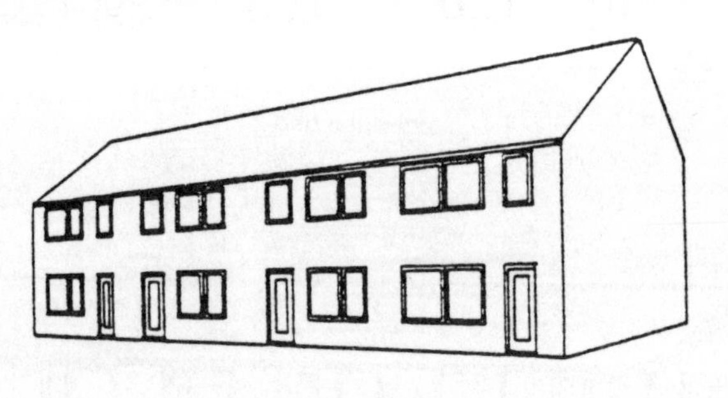

Terraced

Link detached

Houses.

A terrace of town houses.

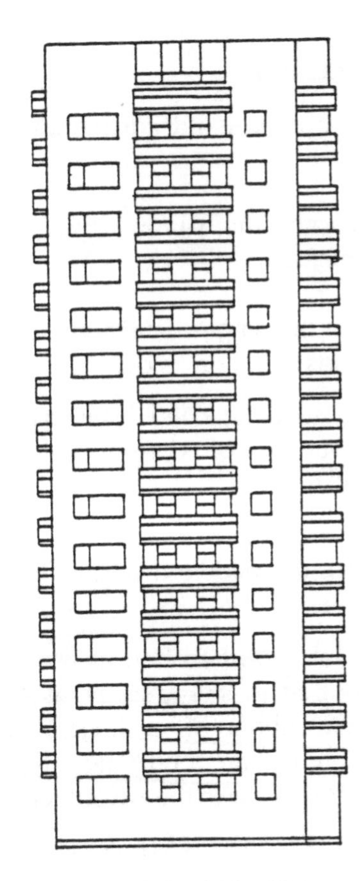

A high rise block of flats.

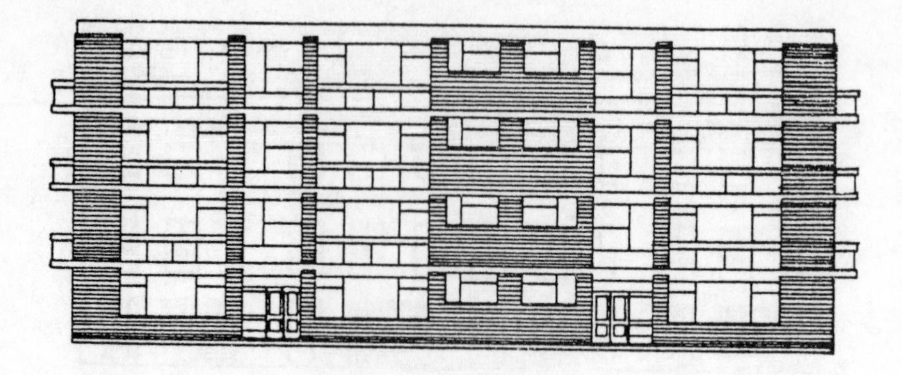

A low rise block of flats.

GROUPING OF DWELLINGS

Dwellings can be constructed as individual units or grouped together to form one building.

Terms used to describe the way in which dwellings, or groups of dwellings, are constructed:

Detached: When a dwelling is detached it is not connected to any other dwelling. It stands alone on its own plot of land.

Semi-detached: A semi-detached dwelling is one of a pair built side by side across the boundary of adjacent plots of land and sharing a common dividing wall.

Terraced: When three or more dwellings are connected together in a row they are terraced, the middle dwellings having two common or party walls.

Link detached: Link detached houses are linked to the adjacent houses by a garage or a single-storey secton of the house, e.g. a porch. They form a continuous ribbon of buildings. Each house is detached, but there is no break in the line of building. The effect is similar to the terraced form of construction with the advantages of a detached house.

Low rise flats: Flats are defined as low rise flats if the block in which they are grouped is four storeys high or less.

High rise flats: Flats are defined as high rise flats if the block, or tower, in which they are grouped is more than four storeys high.

Type of dwelling	*Type of grouping*
Bungalows	Detached. Semi-detached. Terraced.
Houses	Detached. Semi-detached. Terraced. Link detached.
Town houses	Detached. Semi-detached. Terraced.
Flats	Low rise High rise

SOCIAL AND COMMUNITY

Buildings in this area are used for the benefit of the community. They are used to accommodate the facilities society provides for health care, welfare, education and entertainment.

Examples

Social:

- Public houses.
- Clubs.
- Dance halls.
- Cinemas.
- Theatres.
- Football stadia.
- Sports arenas.

Community:

- Hospitals.
- Health centres.
- Schools.
- Railway stations.
- Bus stations.
- Prisons.

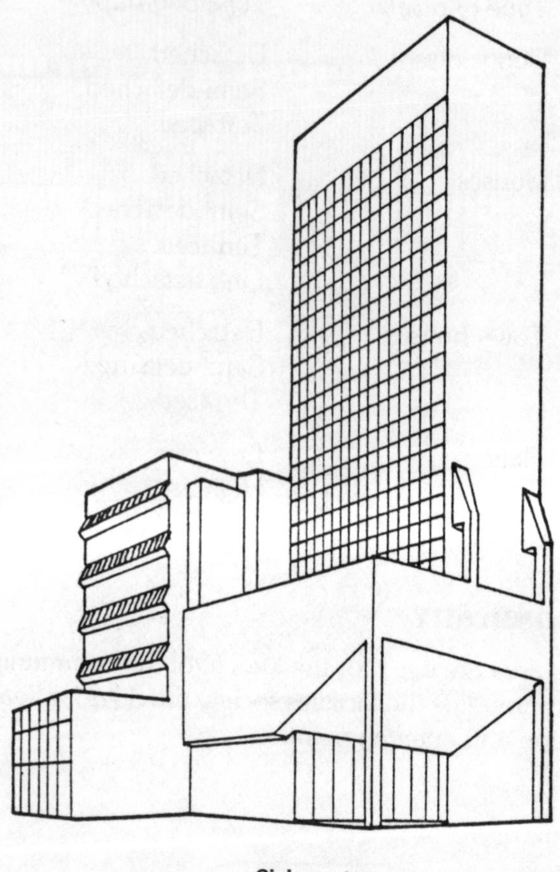

Civic centre.

INDUSTRIAL BUILDINGS

Buildings constructed for industrial use vary a great deal. Their physical size and type of construction depend on the type of business being conducted and the size of the firm.

Examples

Steel making.
Car assembly.
Electronics assembly.
Silicon chip manufacture.
Cabinet making.
Dressmaking.

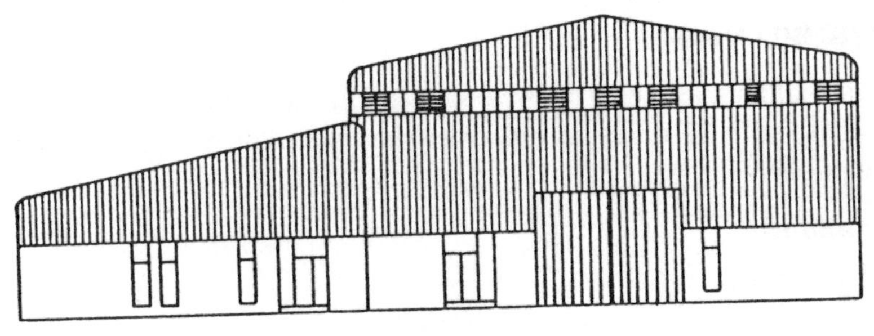

Factory.

COMMERCIAL BUILDINGS

Commercial buildings are where merchandise is sold and services are provided.

Examples

Shops.
Department stores.
Hypermarkets/supermarkets.
Warehouses.
Garages.
Offices.
Building societies.
Insurance offices.
Banks.

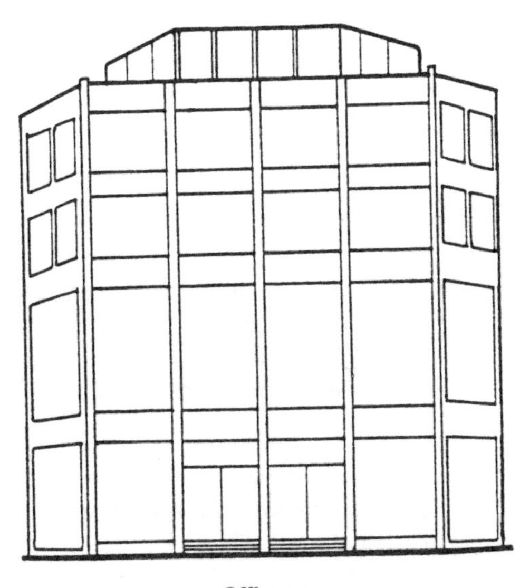

Offices.

CIVIC BUILDINGS

Civic buildings are used by the organisations that enforce law and order, exercise administrative control over society and provide information to the public.

Examples

Town halls.
Law courts.
Register offices.
Public records offices.
Libraries.

CLASSIFICATION OF BUILDINGS BY PURPOSE OR USE

The Building Regulations classify buildings into eight different purpose groups to allow specific regulatory controls to be placed on the different types and uses of buildings.

Purpose group	The purpose for which a building or a compartment within a building is intended to be used
Residential group	
Dwelling house	A private dwelling house.
Flat (maisonette)	A self contained dwelling which is not a private dwelling house.
Institutional	Hospitals, schools and homes used as living accommodation for people who are suffering from disabilities due to illness, old age, physical and mental disability, or for those under five years of age.
Other residential	Hotels, hostels, boarding houses and any other residential accommodation not described above.

Purpose group	The purpose for which a building or a compartment within a building is intended to be used
Non-residential group	
Assembly	Public buildings, or places of assembly where people meet for social, recreational or business purposes (but not office, shop or industrial use).
Office	Premises used for the purpose of administration, clerical work (including writing, filing, book-keeping, typing, drawing etc.), handling money or telephone and telegraph operations.
Shop	Premises used for the carrying on of retail trade or business (including restaurants, public houses, wine bars, cafes, auction rooms, hairdressers etc.) and establishments that hire and repair goods such as videos and tools.
Industrial	Premises as defined in section 175 of the Factories Act 1961 and not including slaughterhouses.
Other non-residential	Places used for the deposit of goods, storage of materials and the parking of vehicles. Premises of a non-residential nature not described above.

2 Structural Forms of Buildings

TYPES OF STRUCTURE

There are three basic forms of structure used in the construction of buildings. Each of these structural forms has many variations which have evolved gradually over the years, being modified and adapted to obtain the maximum benefits from traditional and new materials.

Forms of structure

Masswall.
Framed.
Shell.

Masswall structures

A masswall structure is a solid wall enclosing the total space within the building. It supports the total loading of the building and transmits it to the foundations along the entire length of the wall.

Materials

Bricks.
Blocks.
Stone.
Mass concrete.

Masswall is an easily erected form of structure. It is built up gradually, block by block. When concrete is used as the walling medium, form work has to be erected to create a mould for the liquid concrete. There are two types of masswall structure:

Cellular.
Crosswall.

CELLULAR

The structure consists of walls each joined to its neighbour. The external walls form the boundaries of the building and the internal walls divide the building up into cells making the building cellular. Some of these internal walls are loadbearing walls. Consequently, the external walls do not support the total loading of the building.

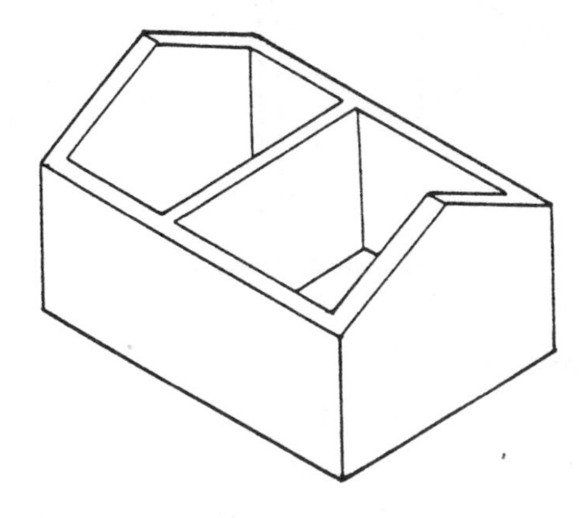

Cellular structure.

CROSSWALL

The crosswall structure is a set of walls parallel to each other, set across the building. These walls take the main structural loads and transmit them to the foundations. Because the crosswalls are free standing they are unstable. To increase their stability it is often necessary to:

(1) rigidly tie the floors to the walls.
(2) increase the thickness of the walls.
(3) build the ends of the walls in the shape of a T on plan.

Crosswall stuctures can be built up to five storeys high. The walls are usually spaced 6 metres apart to accommodate modular units.

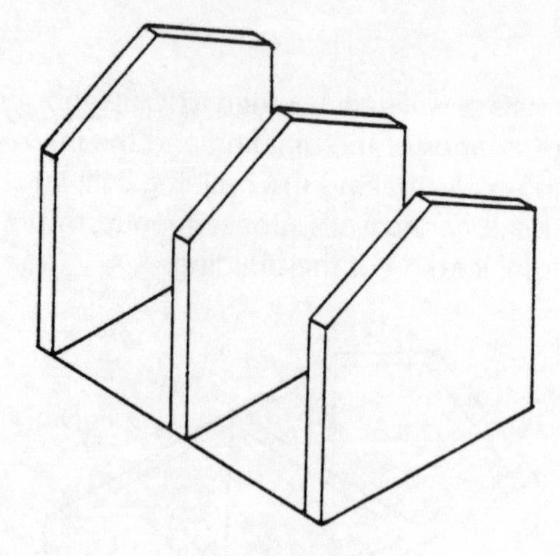

Crosswall structure.

Framed structures

A framed structure is a network of beams and columns joined up to form the skeletal framework of the building. The structural frame carries the total loading of the building and transfers it to the foundations. Cladding is fixed over the framework, or infill panels are placed between its members, to totally enclose the space within the building.

Framed structures are easily erected from pre-made members. The members are simply connected together in the correct sequence to form the structural framework. Problems arise when large members are used and when the height of the structure makes lifting difficult. In these situations a crane will be needed to lift the members into place.

Materials

Timber.
Concrete.
Steel.

Structural frames can be built with pre-made concrete members: this is called pre-cast concrete construction. Most structural concrete frames are cast in situ, i.e. cast in place. In this type of framed construction, form work is needed to make the mould in which the concrete is cast.

There are many different types of structural frame owing to its versatility, and the speed and ease of its erection.

Types

Rectangular frames.
Portal frames.
Triangulated frames.

RECTANGULAR FRAMED STRUCTURES

Rectangular framed structures are a series of upright and horizontal members. They are set at right angles to each other to provide support for the floors, walls and roof. Rectangular framed structures are the most common form of structural frame.

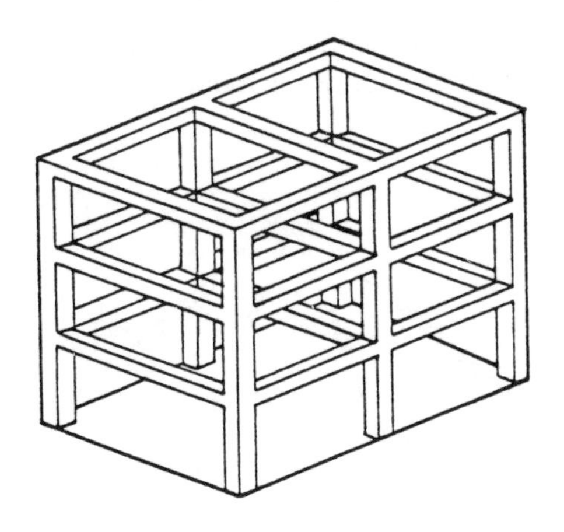

Rectangular frame.

RIGID OR PORTAL FRAMES

Rigid or portal frames are made in a variety of shapes and sizes. Because the structure has no columns, large open areas can be created within the structure. The frames are erected in pairs and need lateral support to stop them falling over. Rigid or portal frame structures are ideal for barns, warehouses, factories and any other buildings that require large open areas.

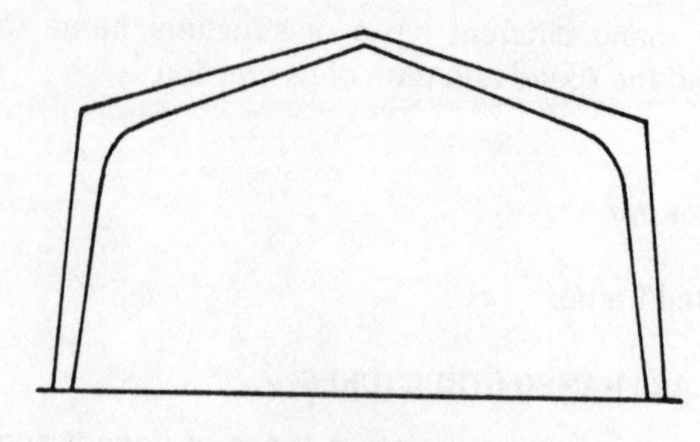

Portal frame.

TRIANGULATED FRAMES

Triangulated frames are the most rigid form of frame possible. Unlike rectangular frames, triangular frames will not distort under load and do not need bracing. The truss, used in the construction of pitched roofs, is one of the most common forms of triangulated frame employed in buildings.

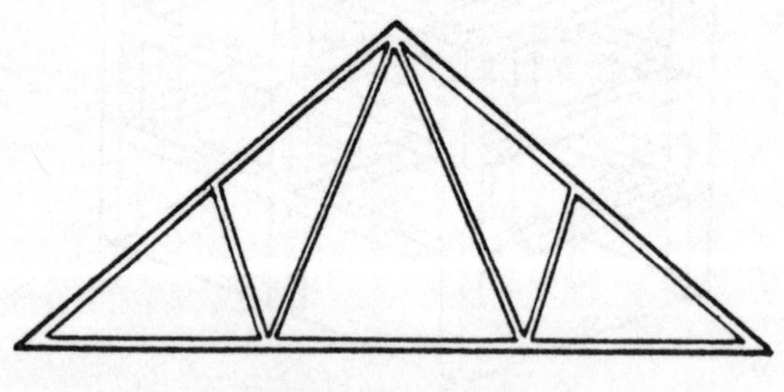

Triangulated frame.

Shell structures

Shell structures are formed by using:

(1) A self supporting rigid skin.
(2) A flexible skin stretched over a lightweight framework.
(3) A flexible skin held up by air pressure.

These create the external surfaces of the building. The surface skin must be strong enough to support its own load and any applied loads.

Many different geometrical shapes and surfaces can be formed with this type of construction, e.g. domes, vaults, saddles and warped segments.

Materials

Steel.
Concrete.
Plastic.
Rubber.
Canvas.

Shell structure.

Structural components

BEAM

The beam is a horizontal member of a frame. It transfers the loads imposed on it, from walls and floors, to its points of support. A beam in its simplest form is a lintel over an opening in a wall.

COLUMN

The column is the vertical member of a frame. It supports the beams and transfers the loads down to the substructure without buckling.

SLAB

A slab is a flat horizontal layer of concrete. It is used to form the floors and roofs of buildings. It can be self supporting, or be supported by beams.

BRACE

A brace may be a strut or a tie. It changes a rectangular frame into a triangulated one. The brace resists sideways forces that could distort the frame out of shape. It is also used to stiffen up the structure.

STRUT

A strut is a brace subjected to compressive loads, i.e. being squashed.

TIE

A tie is a brace subjected to tensile loads, i.e. being pulled apart.

3 Building Elements and Components

ELEMENTS AND COMPONENTS

The structure of a building consists of a number of different elements. In its turn each element is made up from any number of different components. The variety of choice, when designing elements and their component parts, gives buildings their individuality and character.

Elements

There are two kinds of element:

Primary elements.
Secondary elements.

Primary elements – these are the essential structural elements of a building:

Foundations.
Walls.
Floors.
Roof.

Secondary elements are the non-essential parts of a building. They are used to improve the standard of construction, facilities, and appearance of the building:

Services below ground.
Partition walls.
Weatherings.
Cladding.
Services.

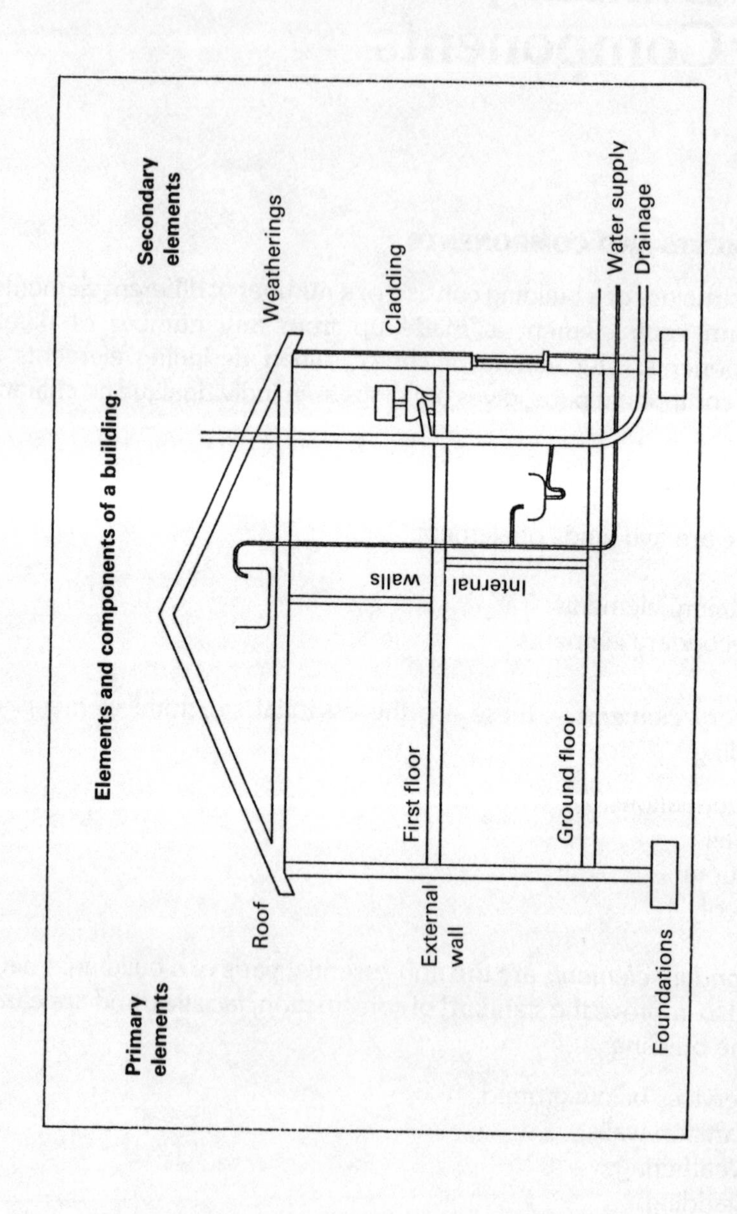

Elements and components of a building.

Components

Components are parts of a whole. Elements can be made up from a wide range of different component parts. For example:

Wall A A house wall

Components:
- Bricks.
- Blocks.
- Brick ties.
- Lintels.
- Window frames.

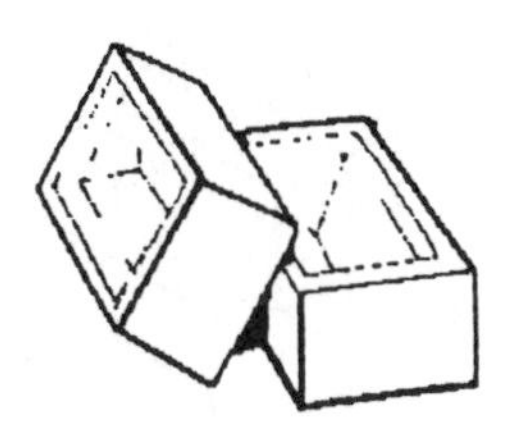

Facing bricks.

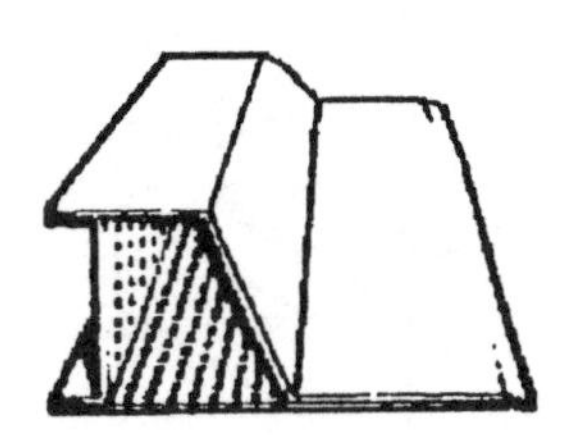

Galvanized steel lintel.

Wall B A garage wall

Components:
- Bricks.
- Lintel.
- Garage door and frame.

Wall C An internal wall

Components:
- Head.
- Sole piece.
- Studs.
- Sound insulation.
- Wall boarding.
- Electric cable
- 13 amp sockets.
- Light switches.

THE FUNCTION OF ELEMENTS

Foundations

The function of the foundation is to spread the loads imposed upon it by the superstucture over an area of ground without any undue settlement occurring. There are recommendations concerning the sizes and depths of foundations.

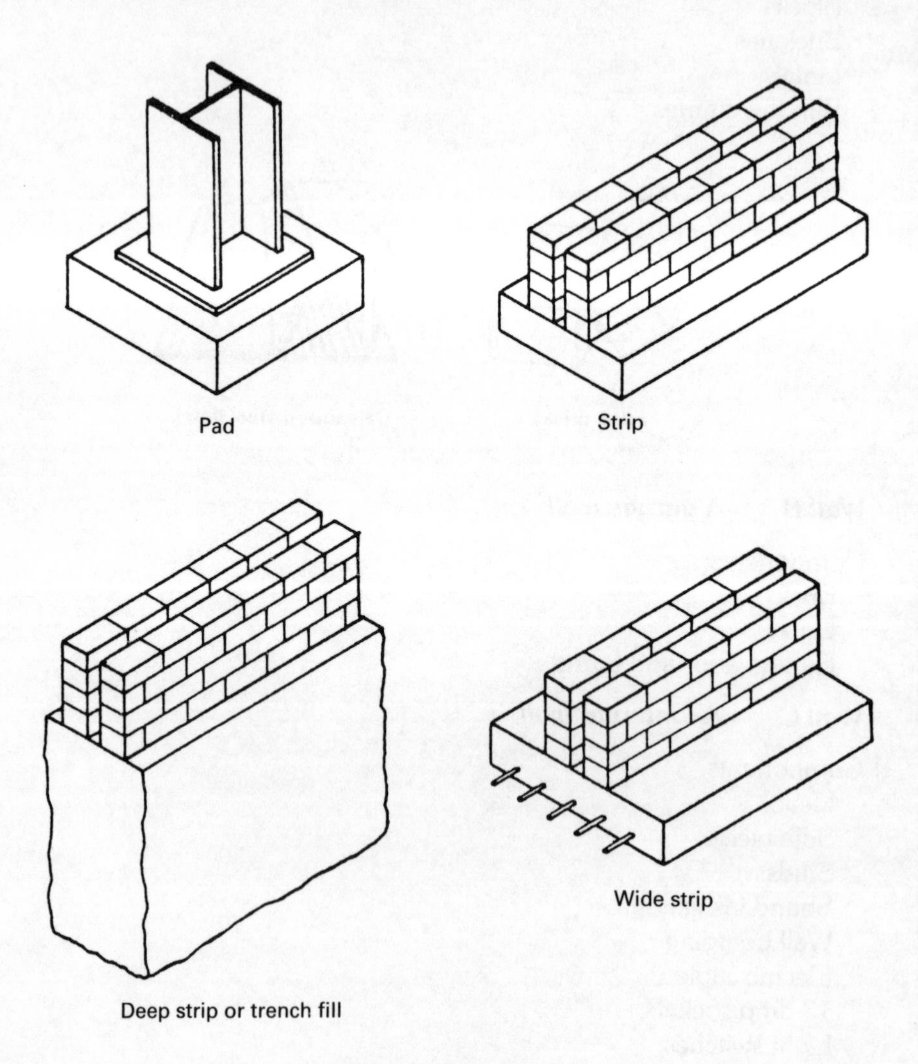

Element: foundations.

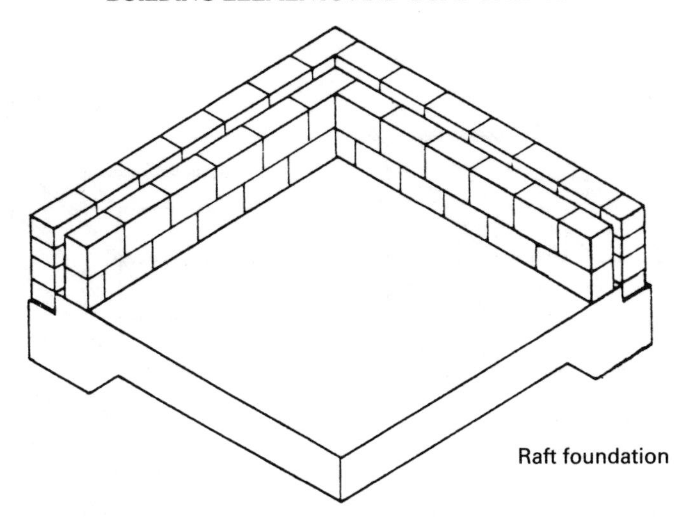

Element: foundations.

The component parts of foundations

Concrete slab.
Reinforcement.
Wall up to DPC level.
Damp proof course (DPC).

Walls

The function of the external loadbearing walls is to:

- Transmit the loads imposed upon them by the floors and roof to the foundations.
- Support any external cladding.
- Separate the internal environment from the external environment.
- Provide insulation between the internal and external environments.

The component parts of walls

Bricks.
Blocks.
Brick ties.
Lintels.
Doors.
Windows.
Thermal insulation.
Internal wall finish.

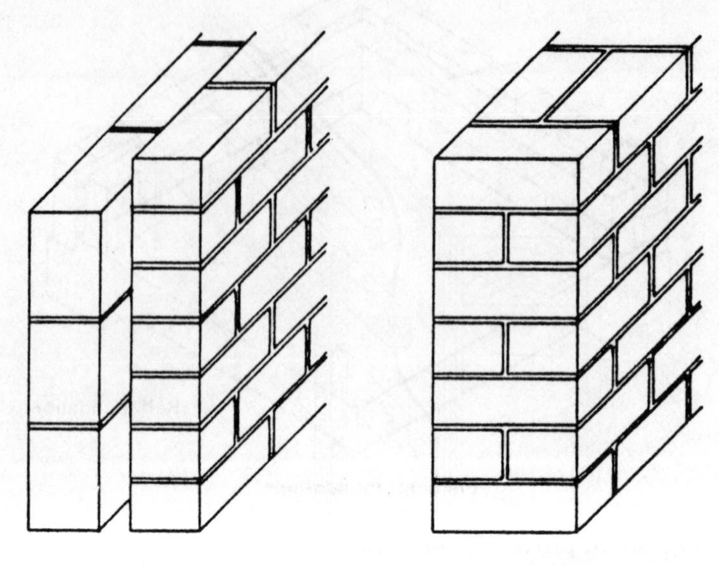

275 mm Brick and block cavity wall. **225 mm Solid brick wall.**

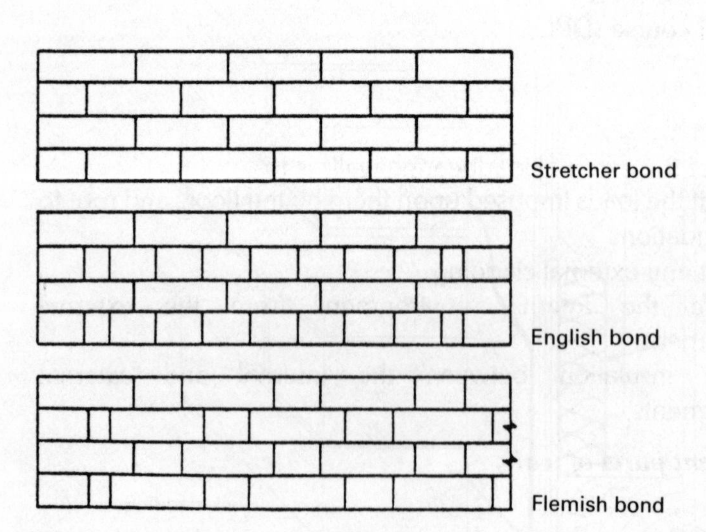

Types of brick bond.

Element: walls.

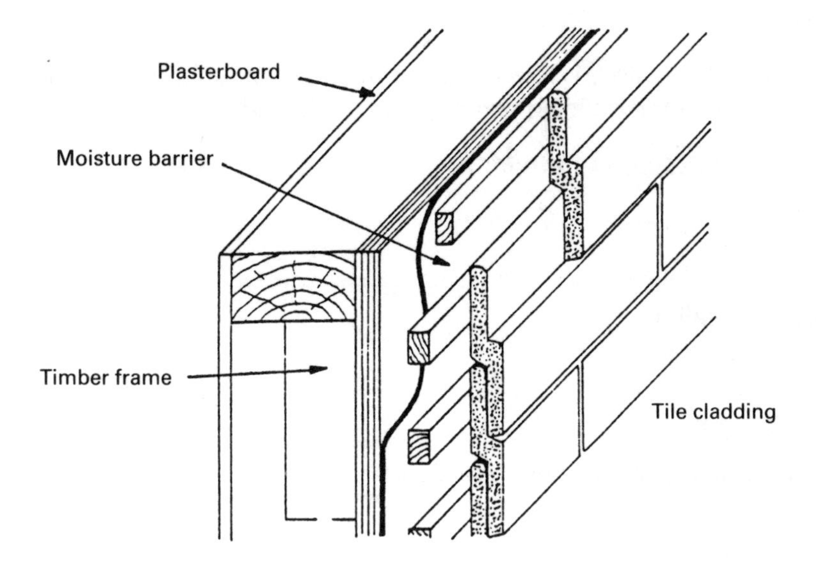

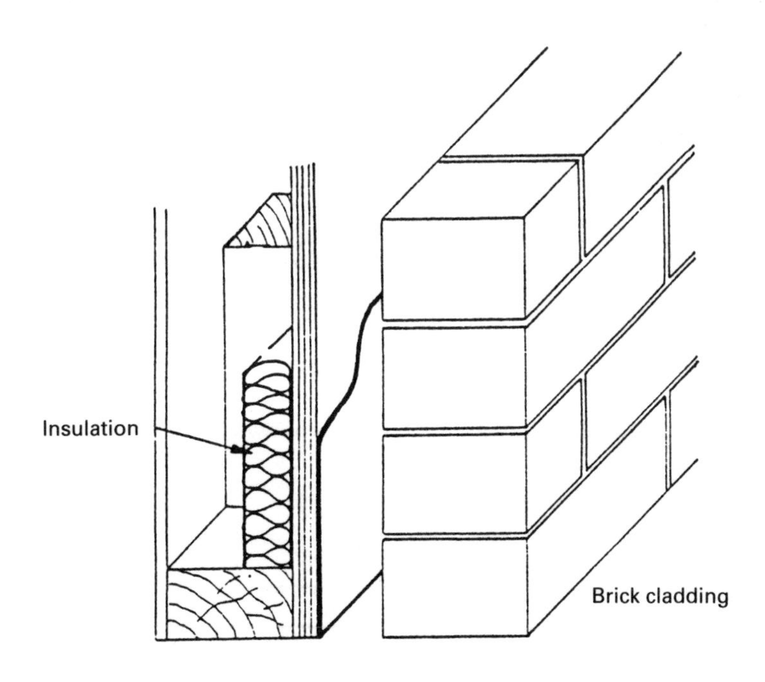

Element: timber framed walls.

Floors

The function of the floor is to provide a firm and level surface on which the occupants of the building can circulate freely and easily.

The component parts of timber floors

Joists.
Floorboards.
Strutting.
Sound insulation.
Ceiling material.
Electric cable.
Ceiling roses.

The component parts of concrete floors

Concrete slab.
Reinforcement.
Floorscreed.
Floor tiles.
Ceiling finish.
Electrical cable.
Ceiling roses.

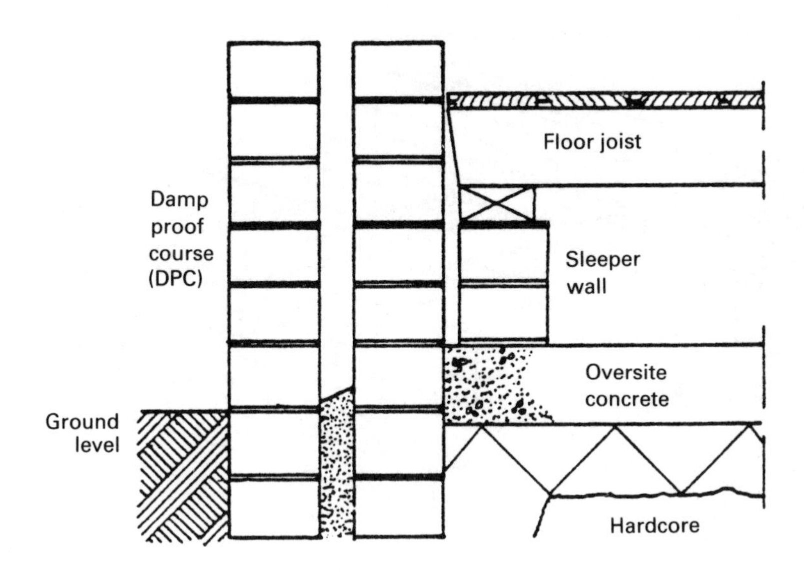

Hollow ground floor.

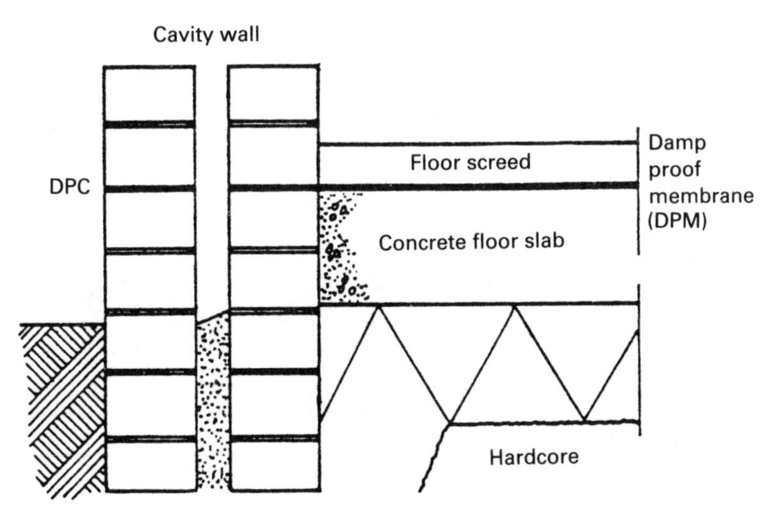

Solid concrete floor.

Element: Floors.

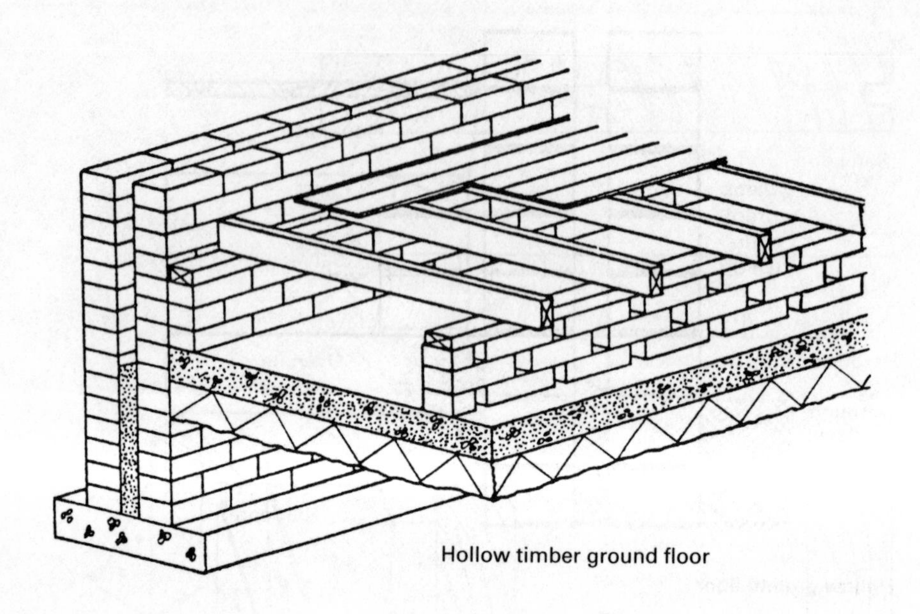

Hollow timber ground floor

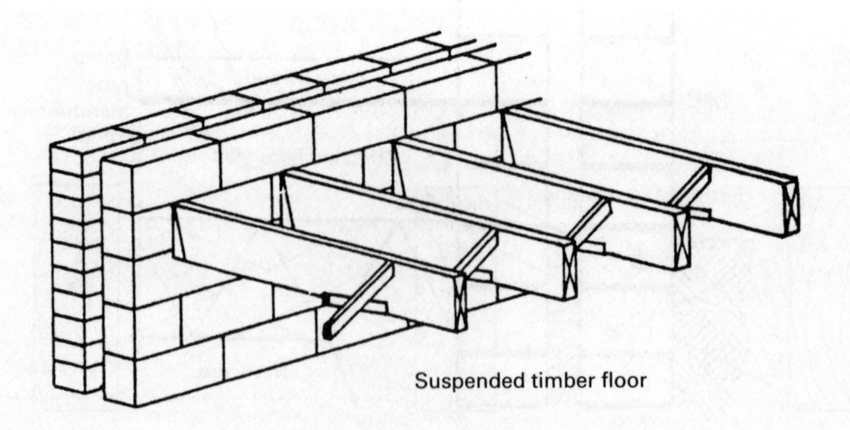

Suspended timber floor

Element: floors.

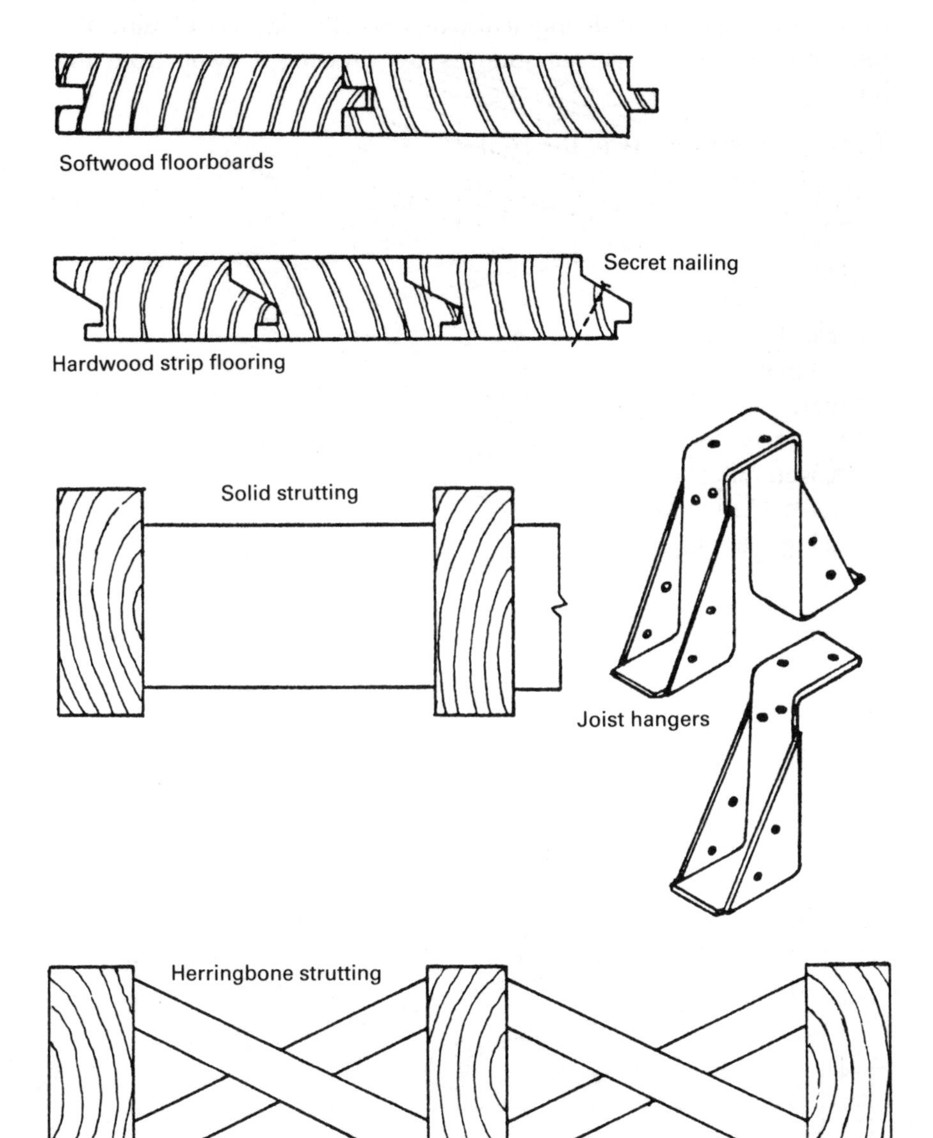

Components: floors.

Roof

The function of the roof is to provide a weatherproof coverng over all the top surfaces of the buildng. It also assists in the disposal of rainwater that collects on its external surfaces by directing it away from the building.

The component parts of the roof

Rafters.
Ridge.
Hips.
Valley rafters.
Jack rafters.
Wall plate.
Purlin.
Roof truss.
Trussed rafter.
Thermal insulation.
Ceiling material.

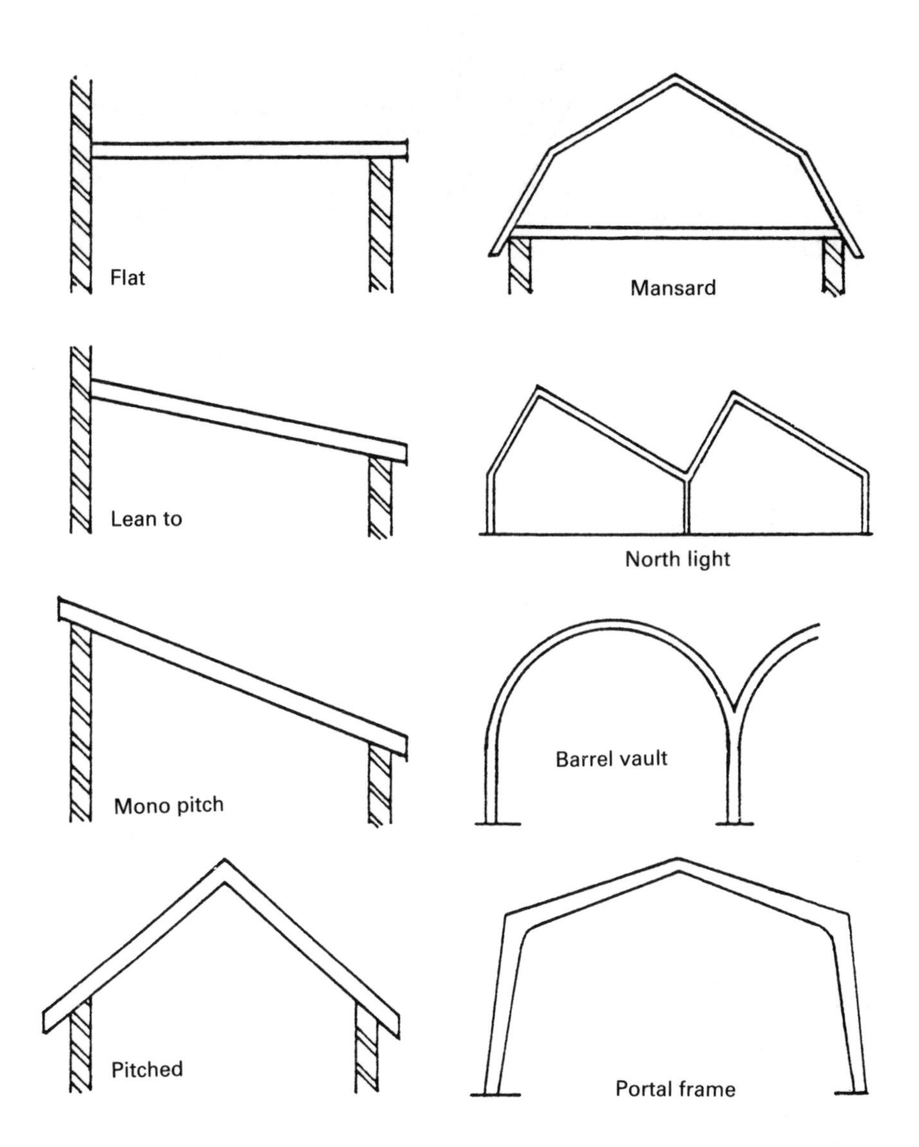

Element: roofs.

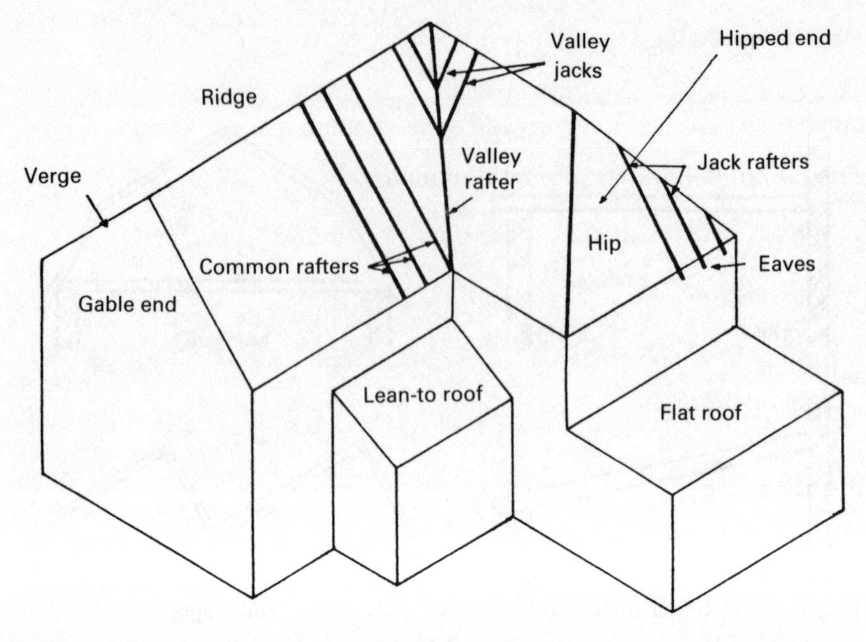

Components: roofs.

Services (below ground)

The function of the services below ground level is to provide a means of supplying water, electricity and, possibly, gas to the building and dispose of its sewage and waste water. The latter is removed from the building and transported to the main drains or sewers in the road.

The component parts of services (below ground)

Drainage

Drain pipes.
Inspection chambers.
Gulleys.
Gratings.

Supply

Pipes.
Cables.
Stop cocks.

Partition walls

The function of the partition walls is to create divisions within the building. These divisions provide areas that are used as living spaces.

The component parts of partition walls

Studs.
Head.
Sole piece.
Noggins.
Insulation.
Wall boarding.

Weatherings

The function of the weatherings on a building is to prevent the entry of moisture and to conduct it quickly away from the building, disposing of it by way of the drainage system.

The components parts of weatherings

Gutters.
Fascia brackets.
Hopper heads.
Down pipes.

Cladding

The function of the cladding is to provide a decorative and weather proof surface over all or part of the external walls of a building.

The component parts of cladding

Supporting framework.
Cladding panels.
Fixings.

Services

The function of the services within a building is the provision of facilities for the occupants. They should enable the occupants to exist in a comfortable, clean, and hygienic environment.

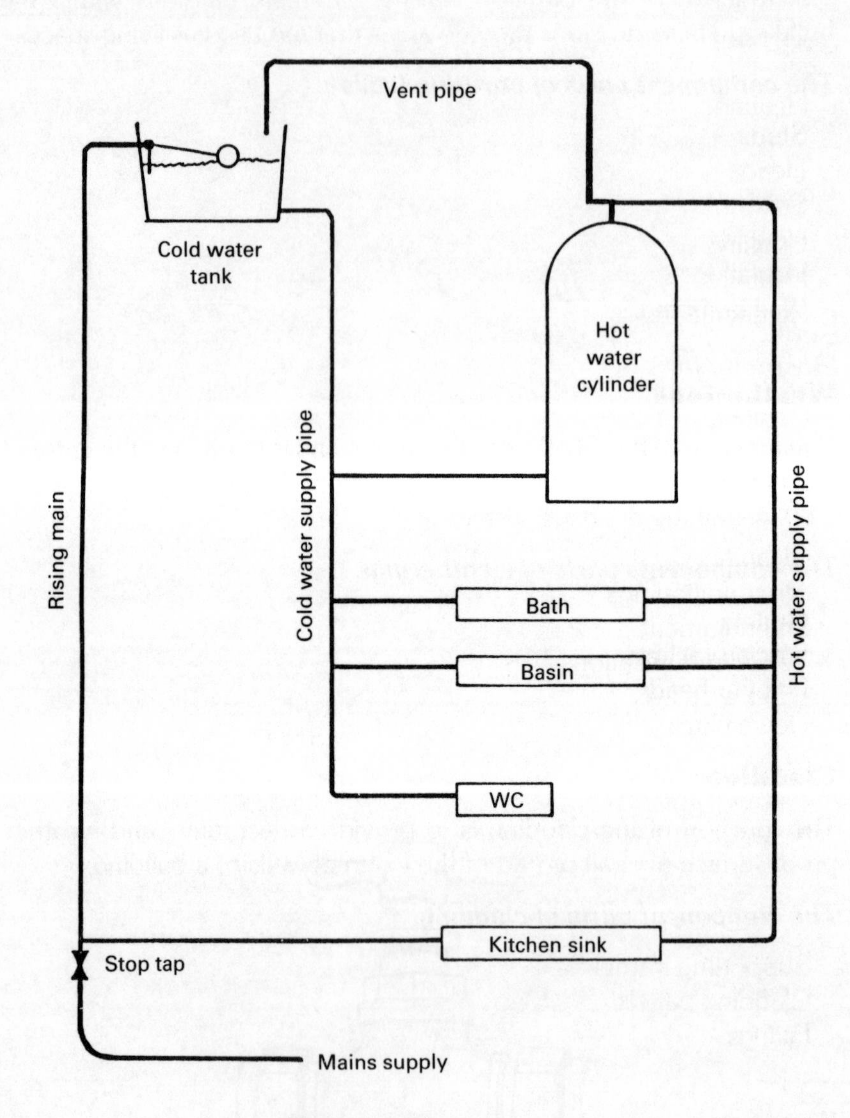

Element: services – water

Water for

Washing.
Drinking.
Cooking.
Heating.
Waste disposal.

Gas for

Cooking.
Heating.
Refrigeration.

Electricity for

Cooking.
Heating.
Lighting.
Refrigeration.
Cleaning.
Air-conditioning.
Entertainment.
Telecommunications.

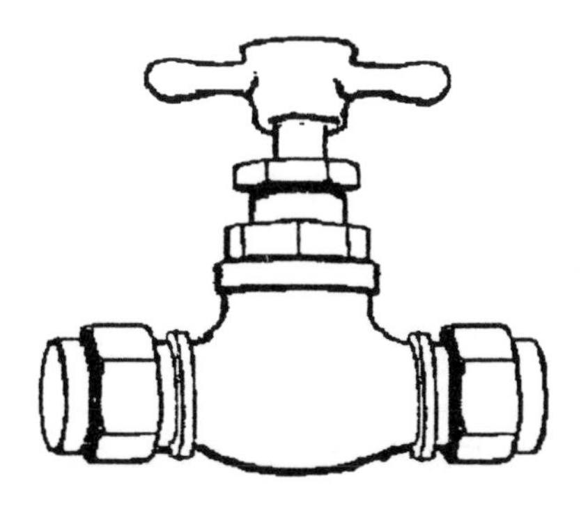

Brass gate valve or stop tap.

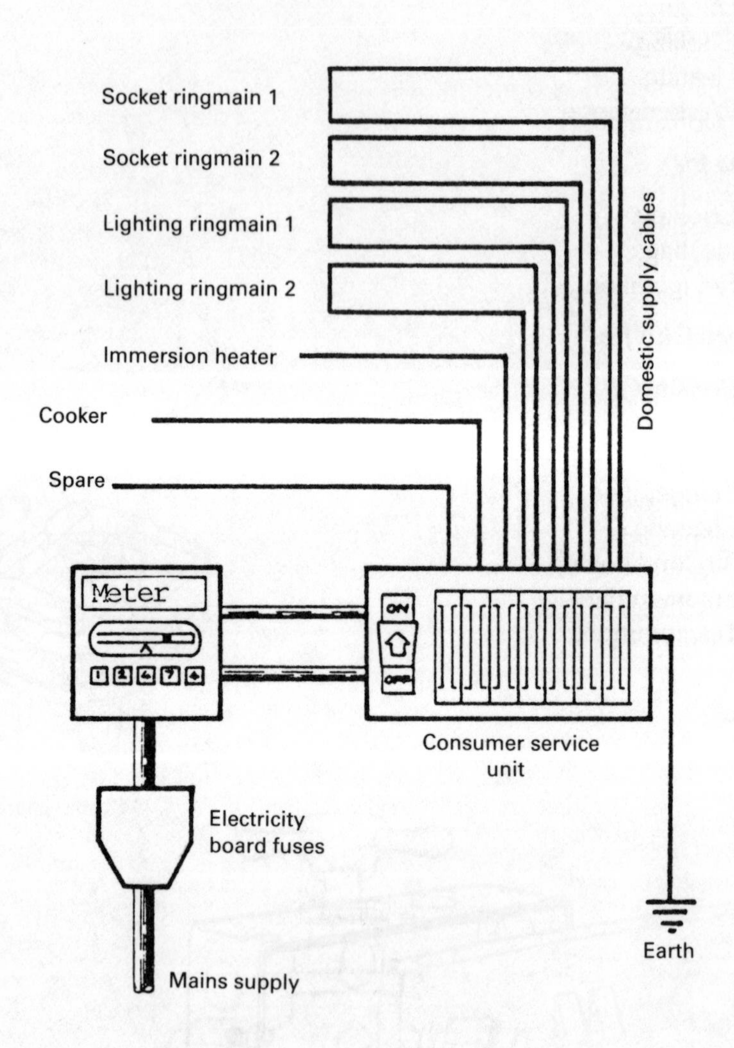

Element: services – electricity.

Drainage for the disposal of sewage and waste water from the building.

The component parts of services

Electricity supply

Electricity Board meter.
Consumer service unit.
Cables.
Junction boxes.
Sockets.
Switches.
Ceiling roses.
Light fittings.

Water supply

Stopcocks.
Pipes.
Fittings.
Pipe clips.
Tank.
Ball Valve.
Taps

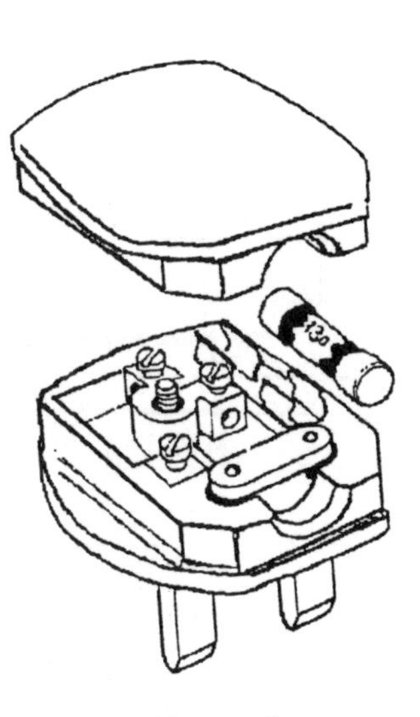

13 amp plug.

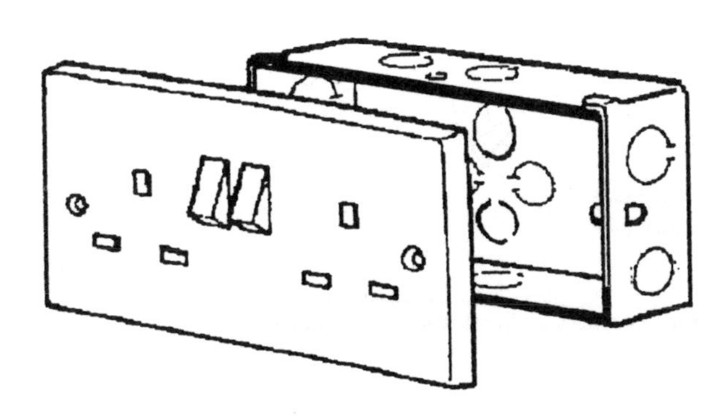

13 amp socket and steel box for flush mounting the socket in a wall.

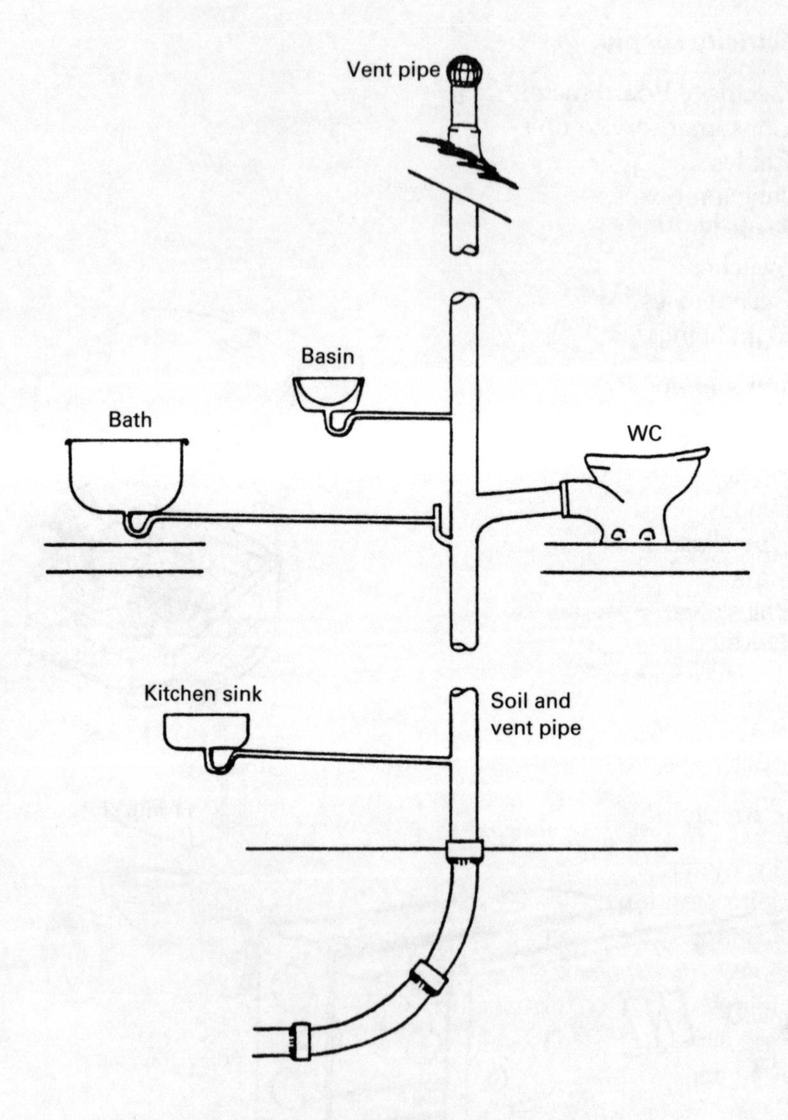

Element: services – single stack drainage.

Hot water supply

Cylinder.
Immersion heater.
Pipes.
Fittings.
Pipe clips.
Stopcocks.
Taps.

Central heating

Boiler.
Tank.
Pump.
Motorized valves.
Radiators.
Pipes.
Thermostat.
Timeclock.
Stopcocks.
Valves.

Drainage

Traps.
Seals.
Pipes.

Gas supply

Gas meter.
Mains stop tap.
Gas taps.
Pipes.
Fittings.
Pipe clips.

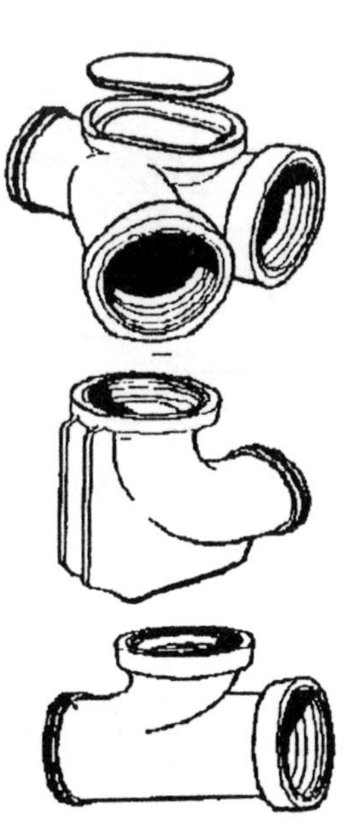

Clay drain pipes.

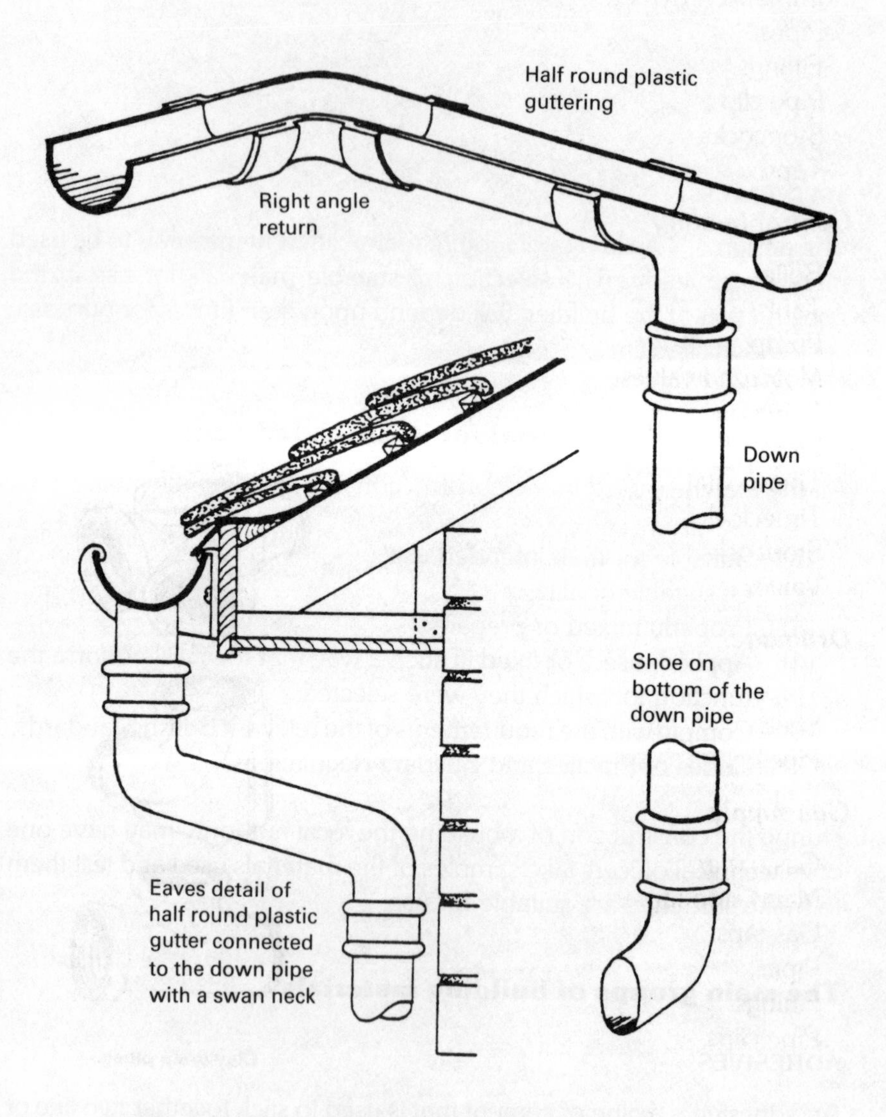

Element: weatherings – rainwater goods

4 Building Materials

MATERIALS

The design of a building will require many different materials to be used in its construction. The selection of suitable materials for use in the construction of the building will depend upon their fitness for purpose, cost and availability.

The fitness of a material for its intended use

All the materials used in the construction of a building must be:

(1) Suitable for their intended use.
(2) Of suitable quality.
(3) Properly mixed or prepared.
(4) Applied, used, or fixed in such a way that they will perform the function for which they were selected.
(5) Comply with the requirements of the relevant British Standards, Codes of Practice and Building Regulations.

During the construction of a building the local authority may have one of its technical officers take samples of the materials used and test them to make sure they are suitable for use.

The main groups of building materials

ADHESIVES

An adhesive is a glue or cement that is used to stick together two like or unlike materials that have rough or smooth surfaces, with a rigid or flexible bond, either permanently or temporarily.

Types

Natural	Animal glue. Casein glue.
Thermoplastic	Polyvinyl Acetate (PVA).
Synthetic resin	Phenol formaldehyde. Urea formaldehyde. Epoxy resin.
Rubber	Contact.

Adhesives.

Uses

Modern adhesives are used to bond together almost anything. They play a major part in the construction of modern buildings, sticking together many items, for example:

Timbers.
Veneers.
Plastic laminated worktops.
Joints in frames.

AGGREGATES

Aggregates are particles or granules of material that are used with a binder to produce a solid mass when set.

By using different aggregates and binders many different kinds of material can be produced.

The characteristics of the material can be altered or changed by the choice of aggregate – the cost can be reduced, the density can be altered, harder wearing surfaces can be created, and different surfaces can be created, and different surface textures, colours and appearances can be created.

Types

Binders	Bitumen. Cement. Plaster.
Aggregates	Sand. Gravel.
Coarse aggregates	Crushed gravel. Crushed stones. Crushed clay bricks.
Lightweight Aggregates	Furnace clinker. Foamed blast furnace slag. Pulverised fuel ash (PFA). Expanded clay, shales or vermiculite.
Heavy aggregates	Iron oxide. Iron. Steel. Lead.

Uses

Sand	Mortars.
Gravel	Concrete.
Crushed gravel	Concrete.
Crushed rock	Mass concrete.
Crushed brick	Hardcore.
Clinker and PFA	Lightweight concrete.
Metal aggregates	Concrete in radiation zones.

BITUMINOUS PRODUCTS

Bituminous products are very resistant to the passage of water, very durable and easy to apply. They are usually black, brown or red in colour.

Types

Bitumen.
Coal Tar.
Pitch.

Uses

Bitumen	Damp-proofing material for DPCs and DPMs. Roofing felts.
Coal Tar	Tarmacadam.
Pitch	Pitch mastic. Adhesives. Used in paint and roofing felt.

BLOCKS

Blocks are larger than bricks and are used for walling or as filler blocks for reinforced concrete floors. They are made from clay, shale or concrete.

Types

Lightweight.
Dense.
Structural / loadbearing.
Non-structural.
Hollow.
Cellular.

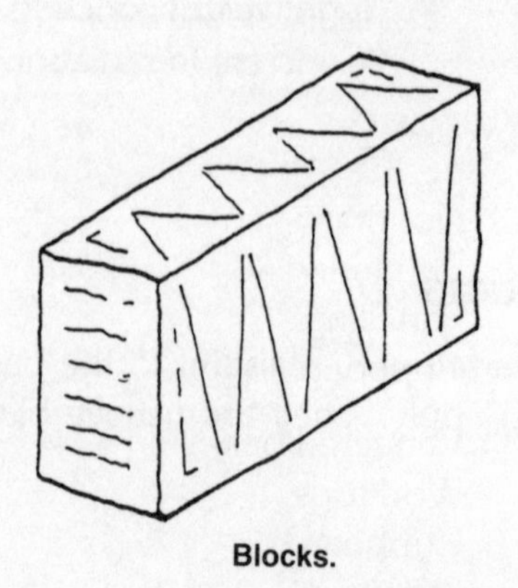

Blocks.

BOARDS, SHEETS AND SLABS

Boards, sheets and slabs are an essential component in the construction of modern buildings. They are available in a wide variety of shapes, sizes, thicknesses and surface finishes. They are easily cut, shaped and bent to cover large areas with few joints. Their strength varies fom the weak fibre boards used for insulation, to the strong such as plywood used for floor decks.

Types

Plywood.
Blockwood and laminboard.
Particle board.
Hardboard.
Fibre board.
Woodwool slabs.
Compressed straw slabs.
Plasterboard.

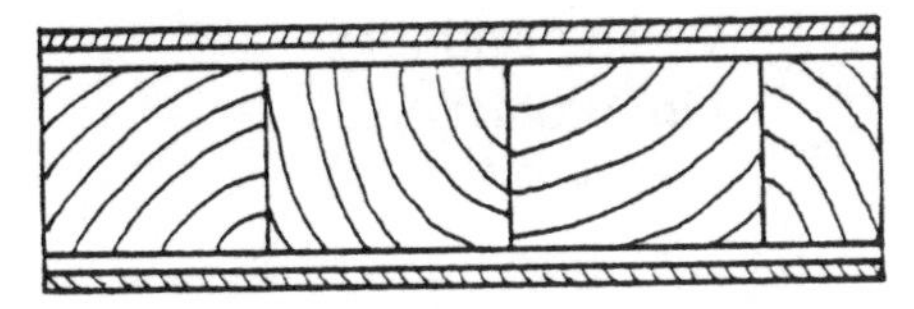

Blockboard

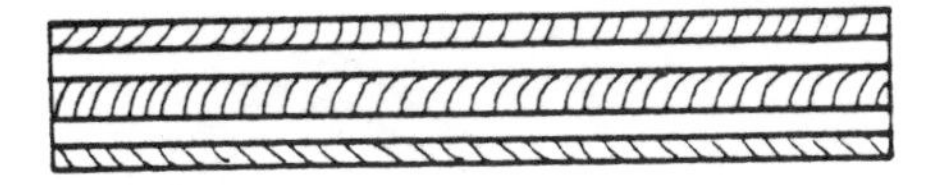

Plywood

Boards, sheets and slabs.

Uses

Plywood	Flooring. Shuttering. Panelling. Cupboards.
Blockboard	Veneered panelling. Worktops. Cupboards.

Laminboard	Veneered panelling.
Particle board	Flooring. Roof decking. Cupboards.
Hardboard	Cladding. Panelling. Flush door panels.
Fibre board	Insulation. Ceilings. Notice boards.
Woodwool slabs	Shuttering. Insulation.
Compressed straw slabs	Partitions.
Plasterboard	Wall cladding. Ceiling cladding. Fire protection.

BRICKS

There are three main materials used in the manufacture of bricks: clay, calcium silicate and concrete.

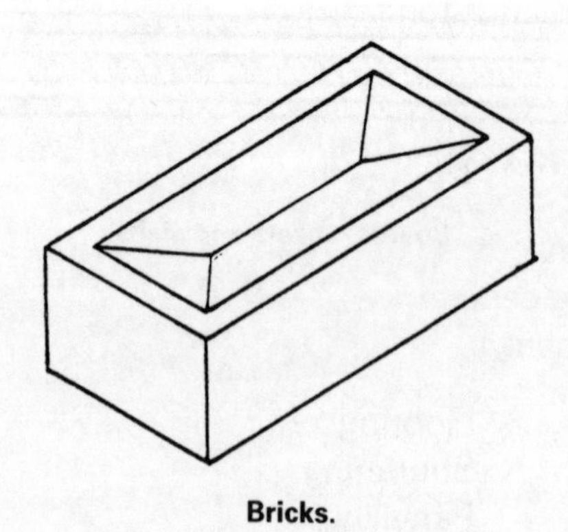

Bricks.

Types

Common Facing Engineering	Clay.

Calcium silicate	Sandlime. Flintlime.
Specials	Concrete. Glass.

Uses

Common bricks	Internal walls.
Facing bricks	Decorative surfaces. Visible surfaces. Features.
Engineering bricks	Structural uses. Damp proof courses.
Special bricks	Fire bricks. Brick paving. Glazed walls.

CEMENT

Cement is a mixture of clay, gypsum and either chalk or limestone. The materials are burnt in a kiln and then ground to a fine powder.

When water is added to cement a chemical action takes place. The result is the gradual stiffening, setting and hardening of the cement.

Cement is often called Portland cement because of its similarity in appearance to Portland stone.

Types

Ordinary Portland cement (OPC).
White cement.
Coloured cement.
Rapid hardening cement.
Waterproof cement.
Masonry cement.

Uses

Concrete.
Floor screeds.
Cement mortar.
Rendering.

CERAMICS

Ceramics are made from minerals such as sand and quartz bonded together with clay and water. The mixture is shaped, dried and then fired in a kiln. The material produced is hard and has a fine smooth surface. Ceramics can be glazed in any colour. The glaze gives the material a very smooth glass-like surface.

Types

Terracotta.
Faience.
Fireclay.
Stoneware.
Earthenware.
Vitreous china.

Uses

Terracotta	Floor tiles. Chimney pots. Air bricks.
Faience	Glazed terracotta.
Fireclay	Firebacks. Refractory bricks. Flue linings.
Stoneware	Drain pipes.
Earthenware	Glazed wall tiles. Sinks.
Vitreous china	Toilets. Basins. Sinks.

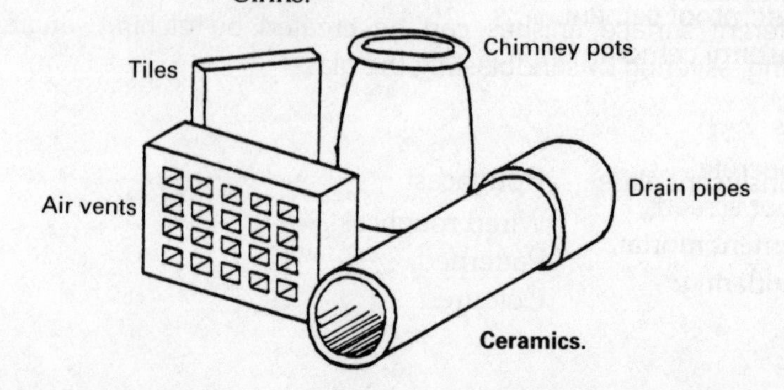

Ceramics.

CONCRETE

Concrete is an artificial rock made from a mixture of coarse aggregates, sand, a cement binder and water. The appearance and properties of concrete are similar to those of limestone rock. The main advantage of using concrete is its versatility. It can be moulded to any required shape and its loadbearing capabilities can be increased by casting in steel reinforcing bars.

Types

Dense concrete.
Lightweight concrete.
Air entrained concrete.

Uses

Dense	Mass concrete. Reinforced concrete. Precast blocks. Concrete pipes. Roofing tiles.
Lightweight	Precast blocks. Precast slabs. Internal partitions.
Air entrained	Thermal insulation.

GLASS

Glass is made by melting a mixture of sand, soda ash, limestone and dolomite in a furnace. When the molton mixture is cooled rapidly it becomes hard and clear.

By adding extra ingredients a wide variety of different glasses can be manufactured.

Different surface finishes can be created by etching, engraving, grinding, silvering or sandblasting the glass.

Types

Translucent glass	Roughcast. Wired roughcast. Patterned. Coloured.

Transparent glass	Sheet. Plate. Polished plate. Float. Wired. Tinted.
Special glass.	Heat resistent. Armour plated. Toughened. Laminated.
Products	Flat glass sheet. Double glazing units. Glass blocks. Corrugated glass. Moulded glass. Mirrors. Leaded lights.

Uses

Window panes.
Roof lights.
Screens.
Protective screens.
Partitions.
Decorative wall panels.
Mirrors.
Exterior wall cladding.

MASTICS

Mastics are used to seal gaps, fill cracks and act as a flexible joint between different materials. To be a good sealant a mastic should not crack, run, blister or harden, and should be weather resistant and stick well to all the surfaces being joined.

Mastics are available in a wide range of colours. They may be applied by hand using a trowel or knife, from a tape or strip, or by squeezing out of a tube by a pressure gun.

Types

Plastic	Bitumen. Butyl rubber.
Elastic	Rubber. Silicone. Polyurethane. Butyl rubber.

Uses

Expansion joints.
Sealing joints between external cladding panels.
Sealing around door and window frames.
Sealing gaps around sinks, baths, showers and lavatory basins.
Waterproof grout for tiles.
Glazing.

METALS

Metals are minerals obtained from metallic ores found in the earth's crust. The metal is extracted from the ore by smelting in a furnace.

Metal can be worked into the required shape in a variety of ways:

Forging – hammering into shape.
Rolling – to produce a continuous strip or sheet.
Pressing – moulding and bending sheets.
Drawing – stretching out into tubes or wires.
Casting – pouring molten metal into moulds.

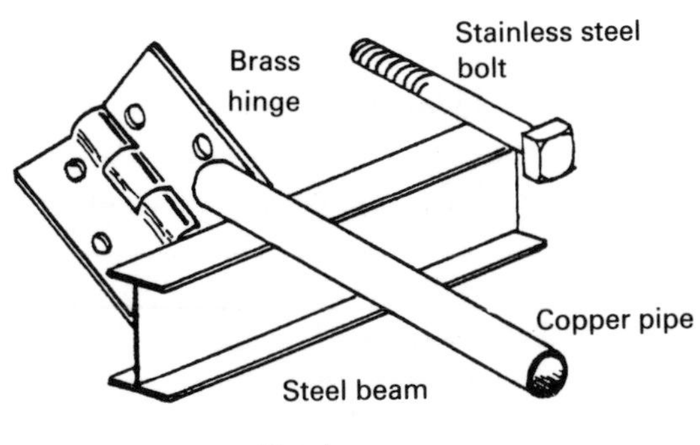

Metals.

Metals can be classified into two groups – ferrous and non-ferrous.

Ferrous metals, i.e. steel and iron, corrode rapidly when exposed to the elements.

Non-ferrous metals, i.e. lead and copper, corrode very slowly forming a thin layer of corrosion over their surface. The layer of corroded metal that forms on their surface creates a protective layer that prevents any further corrosion taking place.

Types

Ferrous metal	Cast iron. Wrought iron. Steel.
Non-ferrous metal	Copper. Brass. Bronze. Nickel. Tin. Chromium. Zinc. Lead. Aluminium. Alloys.

Uses

Cast iron	Manhole covers. Hinges.
Wrought iron	Decorative screens.
Steel	Beams. Brackets. Hinges.
Copper	Pipes. Cisterns.
Brass	Valves. Stopcocks. Decorative fittings. Hinges.
Bronze	Valves. Stopcocks.

Nickel	Plating.
Tin	Plating.
Chromium	Plating.
Zinc	Roof covering.
Lead	Pipes. Roof covering.
Aluminium	Window frame sections.

MORTARS

A mortar is a layer of material more than 3 mm thick used to bond together bricks, blocks, stones etc.

The mortar must have good bonding powers, be durable, easily worked and have minimal shrinkage on setting.

Types

Cement.
Lime.

Uses

Cement	Work below ground level. External walls. Exposed work.
Lime	Internal work. Work with thin joints.

PAINTS

Paint is a very thin decorative and protective coating applied to a surface in liquid or plastic form. The coating dries off forming a hard solid skin.

The type of finish required may need a number of coats of different types of paint or repeated coats of a single paint.

Types of coat

Primer

A protective coat against corrosion or moisture that forms a good base for other paints.

Undercoat	Covers up the material and gives an even solid colour on which finishing coats can be applied.
Finishing coat	Gives the final colour with the required finish i.e. matt, eggshell or gloss.

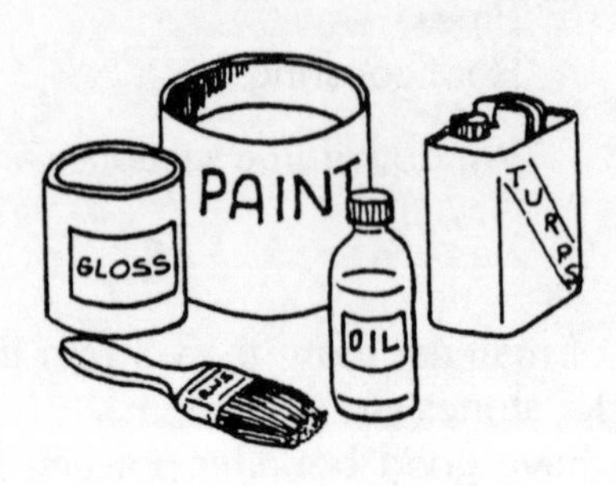

Paint.

Methods of application

Brushing.
Spraying.
Rolling.
Dipping.

Types

Enamel.
Gloss paint.
Undercoat.
Primer.
Oil paint.
Emulsion.

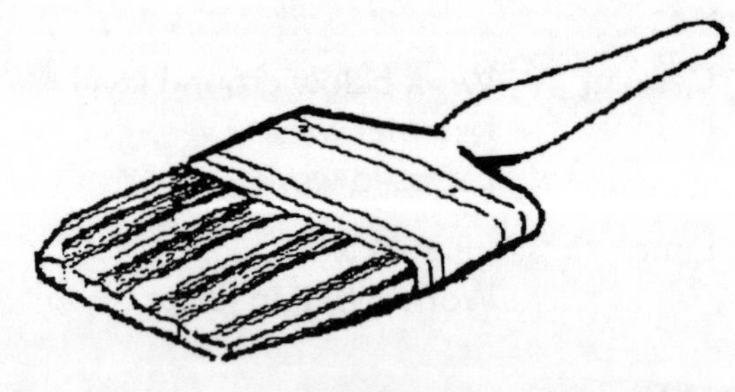
Paint brush.

Uses

Enamel	High gloss finish.
Gloss paint	Gloss finish.
Undercoat	Base coat for gloss.
Primer	Protective sealer for surface.
Oil paint	Internal and external use.
Emulsion	Internal use.

PLASTERS

Plaster is used to provide a smooth surface coat to walls and ceilings on to which a decorative material can be applied.

Plaster is composed of a fine aggregate and a binder mixed with water:

Aggregates	Natural sand. Crushed stone sand. Expanded perlite. Vermiculite.
Binders	Gypsum. Portland cement. Lime.

PLASTICS

Plastics are made from materials obtained from coal or petroleum and become plastic when heated. During its manufacture, whilst it is still soft, the plastic can be moulded into any required shape or form.

Plastics can be divided into two groups – *Thermosetting* and *Thermoplastic*.

Thermosetting plastics once moulded and set cannot be remoulded.

Thermoplastic plastics can be reheated until they are soft and then remoulded into another shape.

Types

Thermosetting	Melamine resins. Phenolic resins. Polyurethane. Urea resins.
Thermoplastic	Acrylic resins. Nylon. Polythene. Polyvinyl chloride. Polypropylene. Polystyrene.

Uses

Thermosetting

Melamine resins	Door furniture. Decorative laminated plastic sheets.

Phenolic resins	Door furniture. Electrical fittings. WC seats. Cellular and foamed products.
Polyurethane	Expanded foam products. Sealants.
Urea resins	Electrical fittings. WC seats.
Thermoplastic	
Acrylic resins	Transparent sheets. Opaque sheets. Coloured sheets. Light fittings. Illuminated signs. Sinks. Baths. Shower cabinets and trays.
Nylon	Door furniture. Hinges. Nuts. Bolts. Curtain track. Sliding door fittings. Castors. Pulleys.
Polythene	Damp proof courses. Damp proof membranes. Bath and sink wastes. Water pipeeesss. Cold water tanks. Floats in cisterns.
Polyvinyl chloride	Soil and waste pipes. Rainwater goods. Electrical conduits. Electrical cable. Insulation. Floor tiles. Corrugated and flat sheets.

Polypropylene	Bath and sink waste. Waste pipes. Water tanks. WC seats.
Polystyrene	Light fittings. Expanded foam products.

STONES

Stones, or rocks, are classified into the following three groups:

Igneous
Sedimentary
Metamorphic

Igneous rocks are formed when molten magma from below the earth's crust cools and solidifies either at the earth's surface or within the crust.

Sedimentary rocks form from layers of accumulated sediment. Some consist of material derived from existing rocks and others consist of the remains of plant or animal life.

Metamorphic rocks are formed when rocks already existing are subjected to intense heat and pressure within the earth's crust which alters their character and appearance.

Types

Geological class	*Group*	*Example*
Igneous	Granite	Shap (pink)
Sedimentary	Sandstone	York stone
	Limestone	Portland stone
Metamorphic	Slate	Blue-grey
	Marble	Serpentino

Uses

Granite	Heavy duty paving. Jetties. Bollards.

Sandstone	Paving. General building use.
Limestone	General building use. Carvings.
Slate	Roofing. Flooring. Wall cladding. Paving.
Marble	Wall cladding. Internal wall lining. Flooring.

TIMBER

Timber is a natural material, organic in composition. There are two main groups of timber: hardwoods and softwoods. Hardwoods are deciduous and have broad leaves. Softwoods are coniferous and have needle-shaped leaves. The difference between a hardwood and a softwood is a botanical distinction. Hardwoods are not necessarily hard and softwoods are not necessarily soft. Pitch pine is a very hard softwood whilst willow, lime and balsa are very soft hardwoods. Balsa is the softest and lightest of all.

Timber, because it is so versatile, is one of the most common materials used in the construction of buildings.

It is durable, strong yet light, easily worked and is available in a wide range of sizes, colours and textures.

Types

Hardwood	Beech. Mahogany. Oak. Teak.
Softwood	Douglas fir. Redwood. Scots pine. Sitka spruce. Western red cedar.

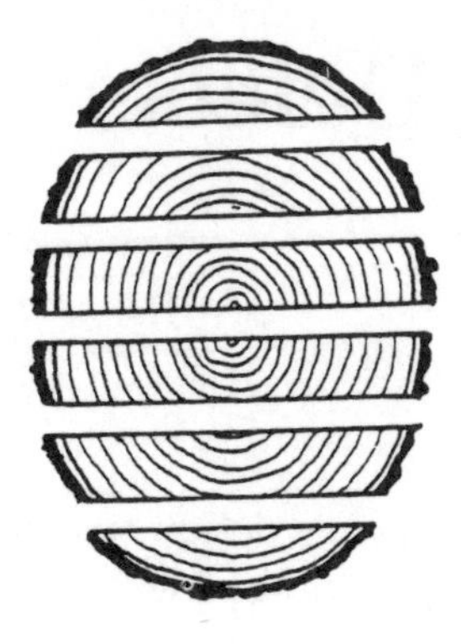

Timber.

Uses

Timber is such a versatile material it has many uses, from heavy structural timbers to fine decorative veneers. The list below is just a small example of its uses.

Hardwood	Decorative timbers. Polished joinery work. Veneers. Structural timbers.
Softwood	General carpentry and joinery work. Structural timbers.

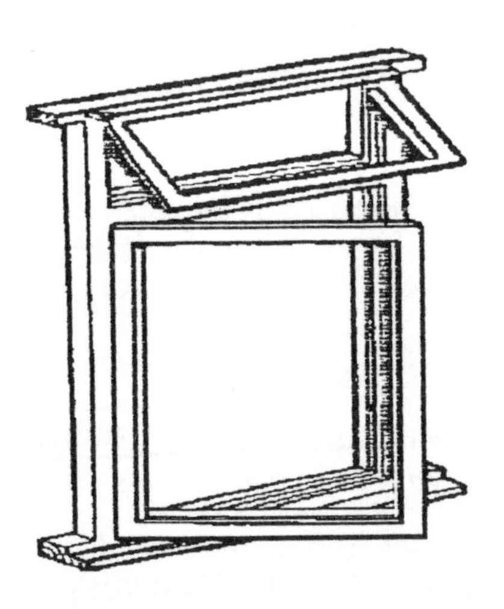

Softwood window frame.

5 Building Defects

When a building fails to meet its design specification, it can be a costly and worrying experience for the builder, its designer and the building owner. Approximately 7 per cent of the work of the construction industry is spent in rectifying defects in buildings.

Defects in the structure of buildings can be attributed to one of two basic causes:

(1) Poor design and construction.
(2) External forces and effects.

Poor design and construction

BADLY DESIGNED

A lack of attention to minor details at the design stage of a building can result in defects being built into a structure.

Example

- Using components and materials in positions for which they were not designed.

POORLY CONSTRUCTED

Poor supervision of the construction process could result in the building not being constructed as it was originally designed.

Examples

- The use of sub-standard materials.
- Putting a smaller quantity of material into the structure than specified, e.g. reinforcing bars in concrete, coats of paint.
- Using cheaper and weaker materials than specified, e.g. weak mortar, cheap paint, poor quality hardwood.

POOR WORKMANSHIP

A lack of quality control and low standards of workmanship will result in defects.

Examples

- Badly fixed components.
- Poorly fitted joints.
- Badly applied mastic seals.
- Poorly applied weatherings.

INADEQUATE MATERIALS

The materials selected for use in a building should perform properly and last for their full life expectancy. They will fail to come up to standard if poorer materials have been substituted. Inadequate fixing of specified materials will also result in the occurrence of defects in the structure.

External forces and effects

External forces can act upon the building and affect the condition of its structure. These forces and effects are a direct result of dampness, movement, chemical, or biological attack. The cause of a defect can often be traced to more than one source.

DAMPNESS

Dampness will cause the rapid deterioration of materials. It is a constant source of trouble to the builder and building owner. The presence of dampness in the structure will establish conditions conducive to other defects e.g. fungal decay and insect attack.

- *Cause of defects:* Rainwater.

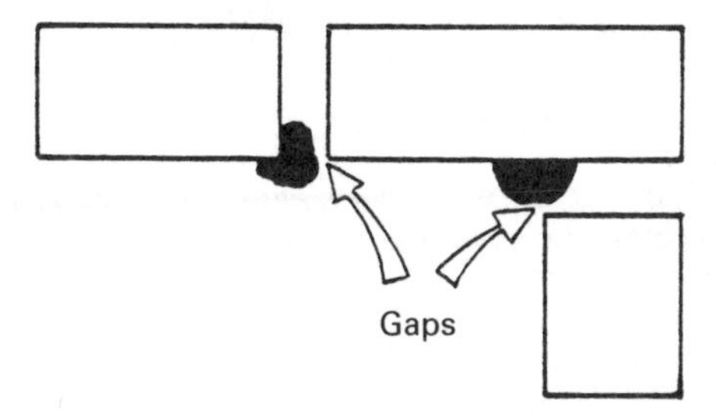

Gaps in joints between materials.

Rain will penetrate the structure of a building through any cracks, joints, gaps and holes that may exist. They may be found in defective components, roof coverings and porous materials, or in the joints between materials.

Defects

Damp patches on walls and ceilings.
Mould.
Defective plaster.
Wet timber.
Fungal attack.

Cure

Prevent moisture entering the structure.
Ensure all mastics and sealants are functioning as intended.
Replace all defective materials.
Replace any defective components.

- *Cause of defects:* Rising damp.

Moisture in the ground is absorbed by the substructure. If there isn't a damp proof course (DPC) in the wall, or a damp proof membrane (DPM) in the concrete floor slab, the dampness will rise above ground level into the superstructure. Rising damp will also occur if the DPC or DPM is defective, i.e. broken or bridged by soil.

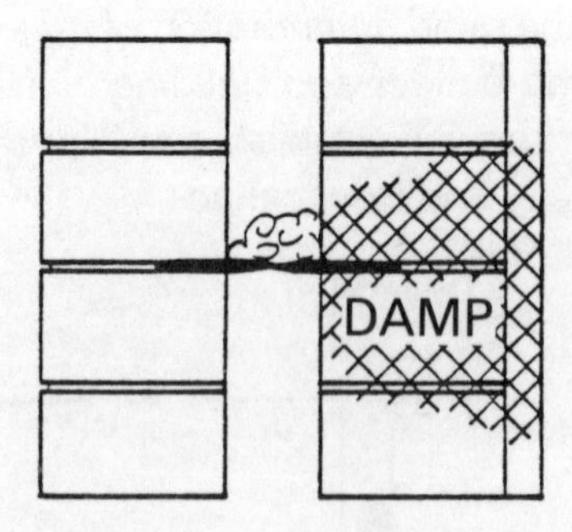

Damp patches on the internal surface of the wall caused by mortar droppings on the brick ties which create a bridge across the cavity.

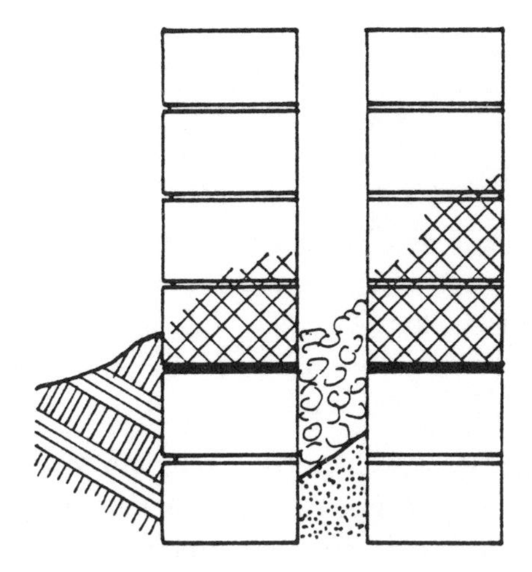

Damp proof course bridged by earth and mortar droppings in the cavity.

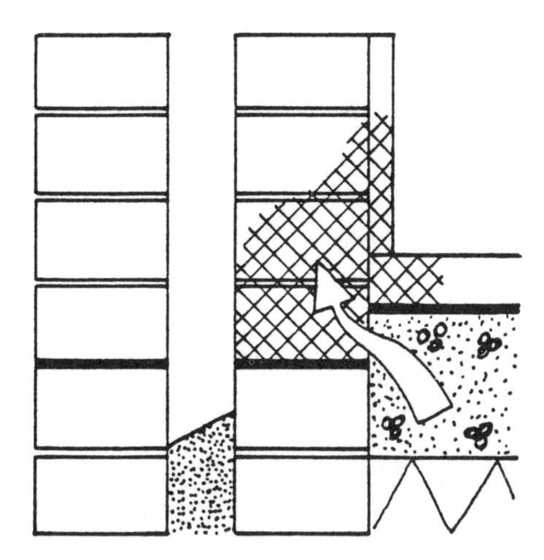

Defective joint between the damp proof course (DPC) and the damp proof membrane (DPM) allowing moisture to rise up into the superstructure.

Defects

Damp patches around the defective areas of concrete floors and walls.
Mould.
Defective plaster.
Wet timbers.
Fungal decay.

Cure

Repair any broken or defective DPCs and DPMs.
Remove any materials that could be bridging the DPC.
Inject silicone DPCs into walls.

- *Cause of defects:* Defective services.

Pipes carrying water can be found inside a building, on its external surfaces and under the ground. Large amounts of water escaping from burst pipes will be easily noticed. If the burst is repaired quickly, damage to the structure should be minimal and easily rectified.

Water seeping from cracked pipes or pipes with leaking joints and seals, is not easily noticed. Damage caused to the structure in this manner can be more harmful. It may occur over a very long period without attracting attention.

Defects

Damp patches.
Mould.
Defective plaster.
Wet timber.
Fungal decay.

Cure

Replace, or repair, any defective gutters, pipes, drains, seals, joints and fittings.

- *Cause of defects:* Condensation.

Warm air can contain large amounts of moisture (water vapour). The warmer the air, the greater the amount of moisture it can contain. When warm moist air cools it has to dispose of its excess moisture. The excess moisture condenses onto the nearest cold surface, covering it with fine droplets of water.

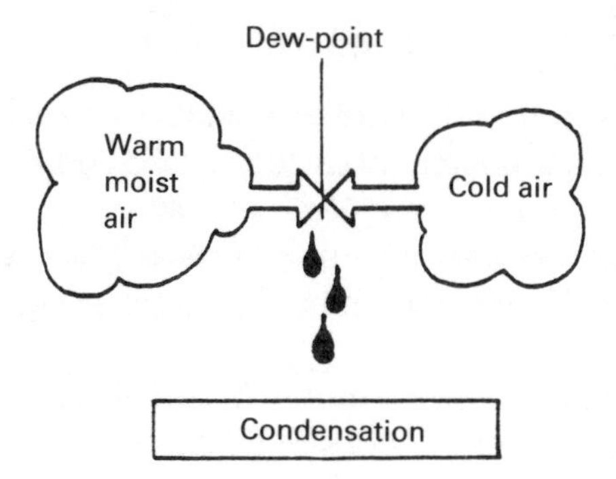

Defects

Condensation on cold surfaces, e.g. external walls, ceramic tiles and window panes.
Mould.
Defective plaster.

Cure

Raise the temperature of the structure by increasing the heating levels.
Provide adequate ventilation to dispose of moist air from the building.
Increase insulation to reduce heat losses and raise the temperatures of cold surfaces.

- *Cause of defects:* Bad maintenance.

Failure to regularly undertake adequate maintenance will create conditions that will allow any of the above failures to occur.

Examples

Blocked gutters and drains.
Rotten timber doors, window frames, and frames caused by defective coatings.

MOVEMENT

Movement in buildings can be of a minor nature, i.e. the simple sticking of doors and windows due to the effects of moisture movement. In extreme cases, movement can take the form of complete structural failure due to ground heave.

- *Cause of defects:* Ground movement.

Movement of the ground supporting a building can affect the stability of the building. Ground movement can be caused by subsidence, landslip, or changes in moisture content of shrinkable soils.

When subsidence occurs the ground sinks to lower levels. It is a direct result of the ground being undermined and collapsing. It often occurs in areas where coal is mined.

Landslip happens when the ground's lateral support is removed or weakened. It happens to the sides of excavations, banks, hillsides and cliffs. If the stability of the ground is reduced, it will slip sideways. The ground's stability can be reduced by changes in moisture content, increased loadings on the ground, soil erosion and undermining.

The ground under and around a building can swell or shrink due to changes in its moisture content. Swelling can be caused by the moisture content of the soil increasing due to broken drains, heavy rain and freezing (frost heave). Shrinkage can be caused by the soil drying out during long periods of dry weather and trees extracting moisture from the ground.

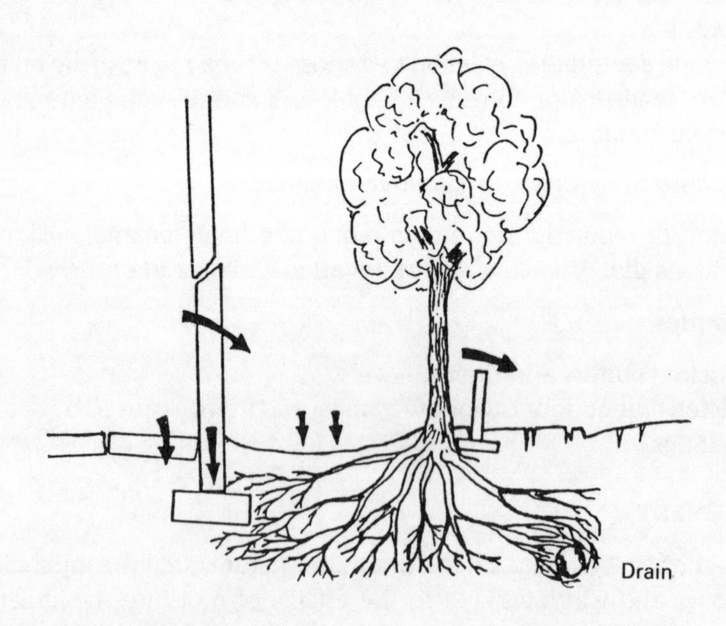

The effects of ground movement caused by removing moisture from the ground.

Defects

Landslip and subsidence which will cause cracking of walls, structural failure and ultimately structural collapse.

Cure

Restrain the ground by using retaining walls.
Strengthen weak or damaged foundations by underpinning.
Remove large trees that are too close to the building.

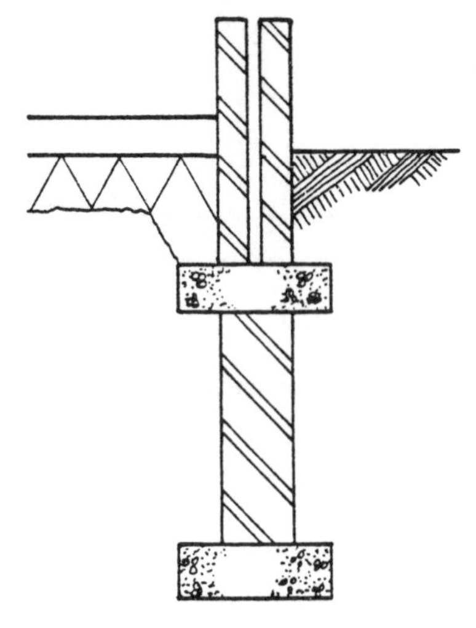

Wall underpinned to transfer the structural loads to a lower level where the ground bearing capacity is higher.

- *Cause of defects:* Overloading the ground.

The loadbearing capacity of the ground can be exceeded by increasing the load that the building applies to it.

Defects

Ground movement causing cracking of walls. structural failure and ultimately structural collapse.

Cure

Remove the source of overloading.
Underpin the structure to obtain a stronger bearing for the foundations.
Remove trees close to the building to reduce the possibility of changes in ground moisture content.

- *Cause of defects:* Overloading the structure.

The structure of a building can become overloaded due to:

Excessive quantities of material being supported by the building.
Structural elements being loaded beyond their design limits.
Weakening structural elements, rendering them incapable of supporting their loads.
Abuse of the structure.

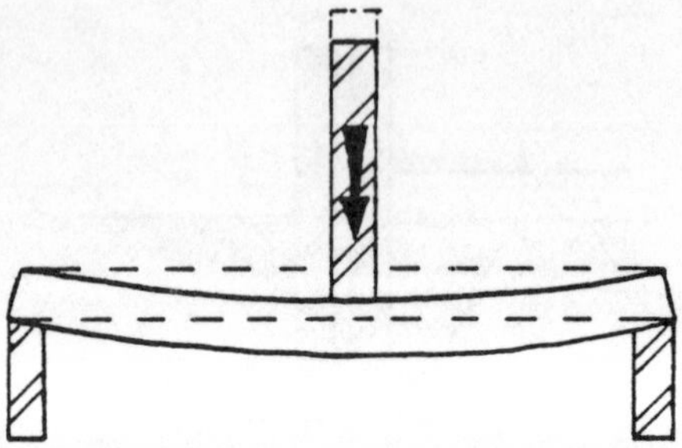

Deflection of beam under excessive load.

Defects

Cracked ceilings.
Sagging floors.
Broken lintels.
Cracked walls.
Major structural failure.
Structural collapse.

Cure

Remove any excessive loads.
Strengthen the structure to a level at which it will support the extra loads.
Return structural elements to their original condition.

- *Cause of defects:* Temperature changes.

Materials expand and contract due to changes in temperature. The building should be designed and constructed to allow materials to expand and contract. Expansion and contraction joints allow materials to move without causing severe cracking of the structure.

Defects

Cracks.
Buckling.
Structural failure.

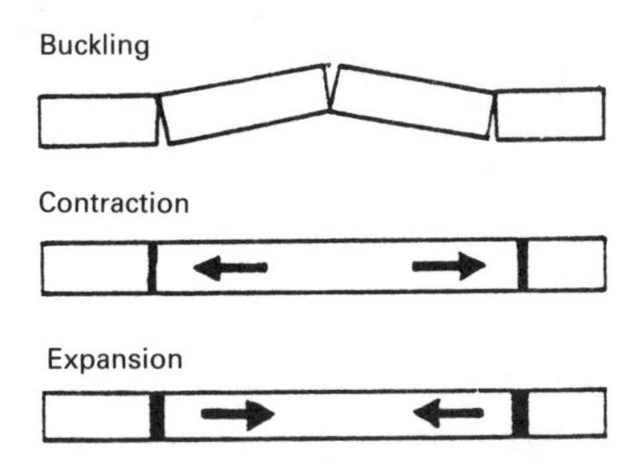

Movement taken up by using flexible joints

Cure

Regular expansion and contraction joints in large areas of concrete and brickwork.

- *Cause of defects:* Moisture.

Materials that are porous, e.g. timber, fibreboard, hardboard, plasterboard, will deteriorate structurally when wet. Increases in moisture content will also deform the shape of materials.

Defects

Swelling.
Warping.
Twisting.
Buckling.
Reduction in strength.
Collapse.

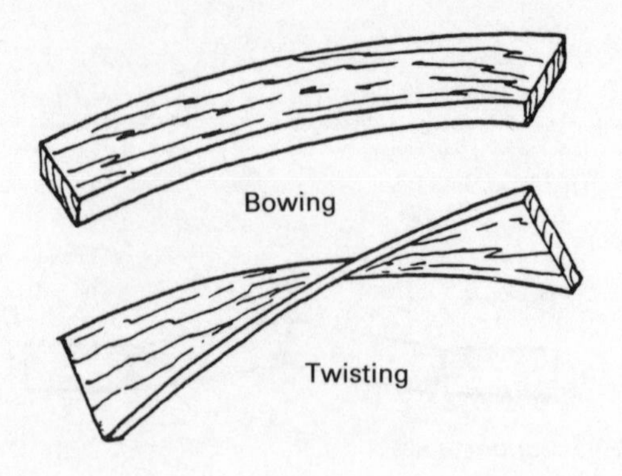

Cure

Keep materials as dry as possible. In the case of exposed timber, coat it with an impervious film of paint, or varnish, to exclude moisture.

CHEMICAL ATTACK

- *Cause of defects:* Chemicals and moisture.

Many materials used in the construction of buildings can be corroded by contact with moisture, or chemicals, in the atmosphere.

The fine dust particles that pollute the air of industrial areas, mixed with rainwater become dilute acids. These dilute acids attack paintwork, corrode metals, and erode the surfaces of brick and stonework.

The corrosion of metals is called oxidation or rusting. Ferrous metals, like iron and steel, will corrode until they are totally destroyed. Non-ferrous metals protect themselves with a thin layer of oxidation. They do not corrode to destruction but remain in good condition for many years.

Defects

Corrosion of metals.
Erosion of stone and brick.

Cure

Use materials that are suitable for the job.
Apply protective coatings.
Use non-ferrous metals.
Remove damaging chemicals from the atmosphere.

- *Cause of defects:* Solar radiation.

Heat and changes in temperature can affect the physical structure of materials. Asphalts, roofing felts, plastics and painted surfaces are most vulnerable.

Defects

Structural deterioration of materials.
Materials discolour, soften, blister and become brittle.

Cure

Use the correct type of materials for exposed locations.
Coat exposed materials with protective or reflective coatings.
Cover flat roofing materials with a reflective layer of fine chippings.

BIOLOGICAL ATTACK

- *Cause of defects:* Fungal attack.

Timber decay is caused by fungal attack. Fungi are plants that feed on vegetable matter, in this instance wood. There are many different kinds of wood-rotting fungi found in buildings. The most common are dry rot and wet rot (of which there are a variety of forms). All fungi need the correct conditions to enable growth to occur, oxygen (air), moisture, food (timber) and a suitable temperature (20°C to 30°C).

The fungus grows from a tiny spore (seed) that has floated through the air and settled on the timber. It will remain dormant on the timber until the correct growth conditions occur. When the conditions are suitable, it produces masses of fine tubular strands (hyphae). These strands passs through the timber consuming the food in the wood. When the fungal growth is well established, a pancake-shaped fruiting body is formed. When ripe, the fruiting body emits spores into the air and the cycle continues.

Dry rot

Dry rot is a very infectious fungus that attacks all timbers. To commence growth it requires timber with a moisture content above 20%, and bad ventilation or stagnant air pockets. One of the most common starting places is in the space below a timber ground floor. Once established it can penetrate through brickwork and travel long distances across other materials, such as steel and concrete, to reach new food sources. The structure and strength of the timber is ultimately destroyed by the fungus.

Recognition

Timber becomes light brown in colour.
Cubical cracking along and across the grain.
A musty smell.
Strands are visible on concealed surfaces of timber.
Fruiting bodies appear.

Prevention

Use treated timber on new work.
Apply preservatives to existing timbers.
Keep all timber in the building dry.
Repair all sources of dampness.

Cure

Remove and burn all infected timber.
Remove and burn all timber within 1 metre of the infected area to ensure that none of the fungus remains.
Sterilize all surrounding concrete, brickwork etc, by heating with a blowlamp.
Apply fungicides to the surrounding areas of brickwork and concrete.
Apply preservatives to all existing and any replacement timber.

Wet rot

Wet rot does not normally extend beyond the area of wet timber. It is easy to eradicate and its recurrence is unusual. Exposed timbers in doors and windows often suffer from wet rot if they are not adequately protected.

Recognition

Timber becomes dark brown in colour.
Cracks occur along the grain.

Prevention

Use treated timber on new work.
Apply preservatives to existing timbers wherever possible.
Regularly renew protective coatings on surfaces of timber, e.g. paint on timber window frames.

Cure

Remove sources of moisture.
Replace any defective timber.
Use only treated timbers.
Apply protective coatings e.g. paint.

- *Cause of defects:* Insect attack.

Wood boring beetles are small brown insects that attack timber. They can damage the appearance of the timber and severely reduce its structural strength.

The female beetle lays her eggs in cracks in the timber. When the eggs hatch out, the worm-like grubs (larvae) eat their way into the timber. They feed on the cells of the timber, leaving tunnels behind them. The full grown larva becomes inactive (pupa), whilst it changes into a beetle. When the pupal stage is over the fully grown beetle emerges through a small flight hole on the surface of the timber and flies away.

Recognition

Flight holes on the surface of timber.
Wood dust from flight holes visible.

Prevention

Use treated timber on new work.
Apply preservative to existing timbers.

Cure

Treat all infected and adjacent timbers with preservative.
Inject preservative down flight holes visible on coated surfaces of the timber.
Remove all badly infected timber and burn it.
Replace with new timber treated with preservatives.

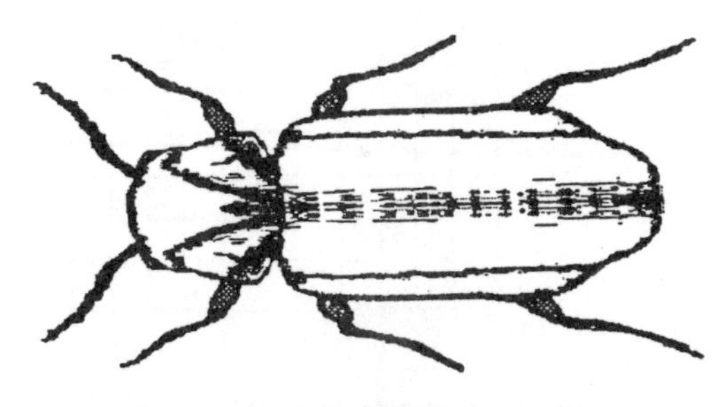

Furniture beetle.

Type of defect / Cause of defect	Dampness				Movement				Chemical		Biological	
	Poor maintenance	Rainwater	Rising damp	Defective services	Ground movement	Structural overloading	Temperature changes	Moisture	Chemical attack	Solar radiation	Fungal decay	Insect attack
Blistering	●			●			●			●		
Blocked drains	●		●	●	●							
Blocked gutters	●			●								
Broken lintels					●	●						
Brittleness				●			●		●	●		
Buckling					●	●	●	●				
Condensation			●									
Corrosion	●		●	●					●			
Cracks					●	●	●	●		●		
Cracks in timber						●					●	
Cracked ceilings					●	●		●				
Cracked walls					●	●	●	●				
Damp patches	●	●	●	●								
Defective plaster	●	●	●									
Deterioration of materials	●	●	●	●			●		●	●	●	●
Discoloration							●		●	●	●	
Wet rot and dry rot		●	●	●							●	
Holes on surfaces of timber												●
Landslip	●				●			●				
Mould	●	●		●								
Reduced strength			●	●	●	●		●	●	●	●	●
Sagging floors					●	●		●				
Shrinkage							●			●		
Softening							●		●	●		
Structural collapse					●	●		●				
Structural failure				●	●	●		●			●	●
Subsidence					●			●				
Swelling	●	●	●	●				●				
Twisting				●								
Warping	●	●		●								
Wet timber	●	●	●	●								
Wood dust												●

Summary of defects and their causes.

6 The Construction Industry

The construction industry is one of the largest sectors of British industry and consists of two distinct areas of operation: building and civil engineering.

Building work involves the construction, maintenance and adaptation of houses, shops, offices, schools and other buildings, as well as large scale projects such as the construction of new towns and shopping complexes.

Civil engineering work is usually large scale and involves the construction and maintenance of dams, power stations, tunnels, motorways, roads, sewage works and the drainage systems that feed them.

Construction firms

The construction industry consists of many different types and sizes of organisations. The backbone of the industry is the large number of small firms that employ fewer than 25 employees. They undertake approximately one third of the work, in value, and employ approximately one third of the industry's workforce. The large firms whilst few in number, undertake approximately half the work, in value, and employ approximately 45% of the industry's workforce.

The economics of construction

Work is undertaken for either private or public sector clients. The amount of work undertaken in each financial year depends a great deal on the financial state of each sector. Levels of finance are influenced by the state of the country's economy. When there is an ecomomic boom, or when government and local authorities are willing to invest in construction, the amount of work becomes plentiful and output grows rapidly. During times of economic stagnation the level of work drops and the industry slows down. A term used to describe the up and down effect on the quantity of work available is STOP-GO.

Other factors also have an effect on the costs and productivity of the industry. When there are increased costs of labour and materials, due to scarcity and constantly rising prices, clients may be discouraged from investing their money in construction work.

Construction employment

Variations in labour costs are caused partly by an inadequate supply of trained operatives. In periods when there are low levels of work the construction industry employs fewer trainees. Consequently, when work levels increase there are fewer trained operatives to fill the vacancies. The labour market then becomes subject to the pressures of supply and demand. Increased demands on a small supply of labour creates the conditions for rising wage rates.

To counterbalance the stop-go sequence of employment and training, a variety of training schemes has been used. The Construction Industry Training Board (CITB) provides grant aided Youth Training Schemes for trainees. When conditions have been right the CITB has also been engaged in Counter Cyclical Training, in an attempt to train people when demand for workers is at a reduced level. In this manner a pool of fully trained operatives would be created ready for when work levels increase.

A further difficulty arises because of the way in which the industry employs its workers. A large number are self employed, which results in reduced levels of training and fewer training places, lower safety standards and increased accident figures. The construction industry accounts for nearly 70% of all work-related injuries to self employed people.

The structure of construction firms

Construction firms range in size from small one man operations to very large multi-national organisations. Between them they undertake a wide range of work, each firm often specializing in one particular type of project or activity. Because of the different types of work undertaken by each firm, each one tends to have its own organisation structure.

Organisation structures

An organisation structure establishes the relationship between the different members and parts of an organisation, establishing how they work together and co-ordinate with each other. It also shows the levels of seniority and the chain of command.

A properly designed organisation structure will enable a firm to function effectively and efficiently.

The organisation structures given as examples are for:

- a large construction firm.
- a large building site.
- a small general builder.

Information about the people involved in the organisation structures can be found in greater detail in the next chapter.

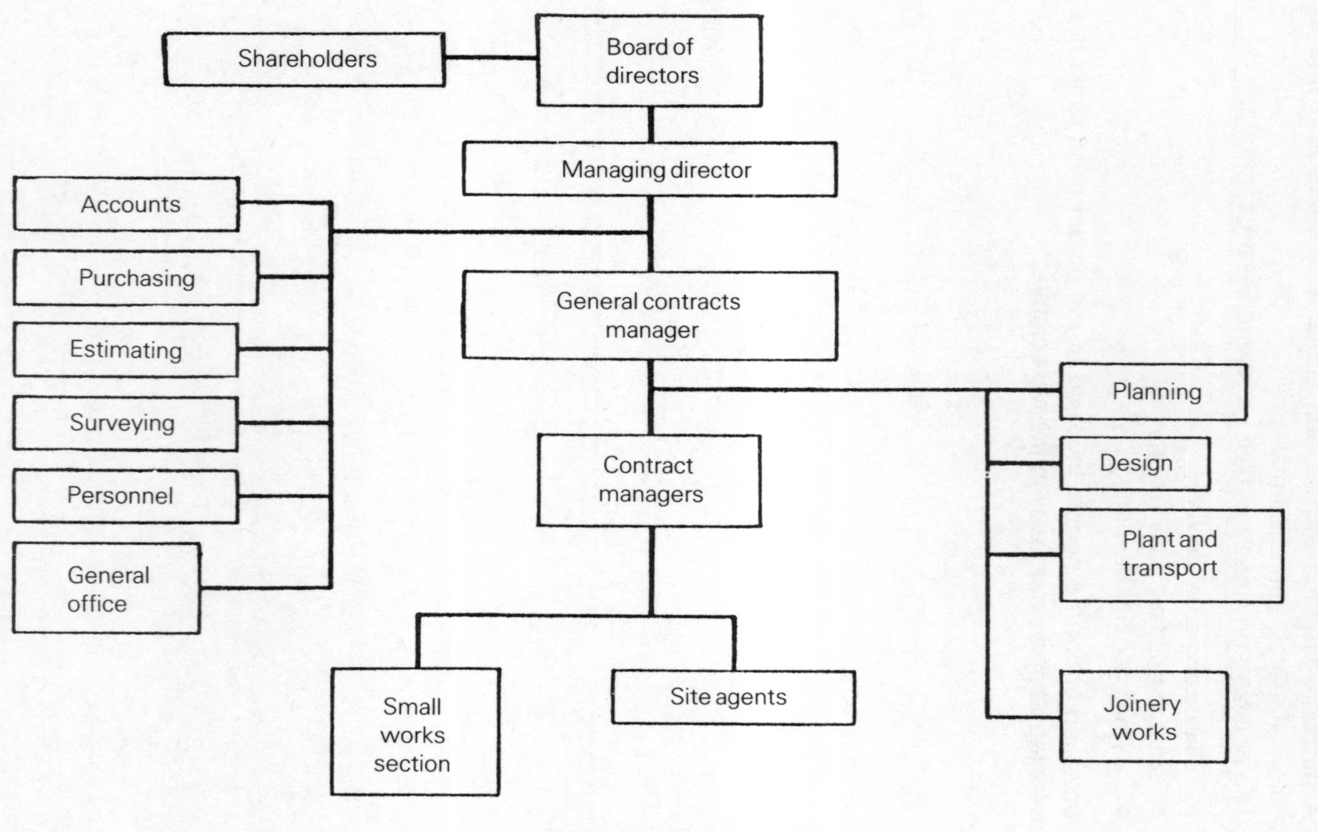

Organisation structure of a large firm

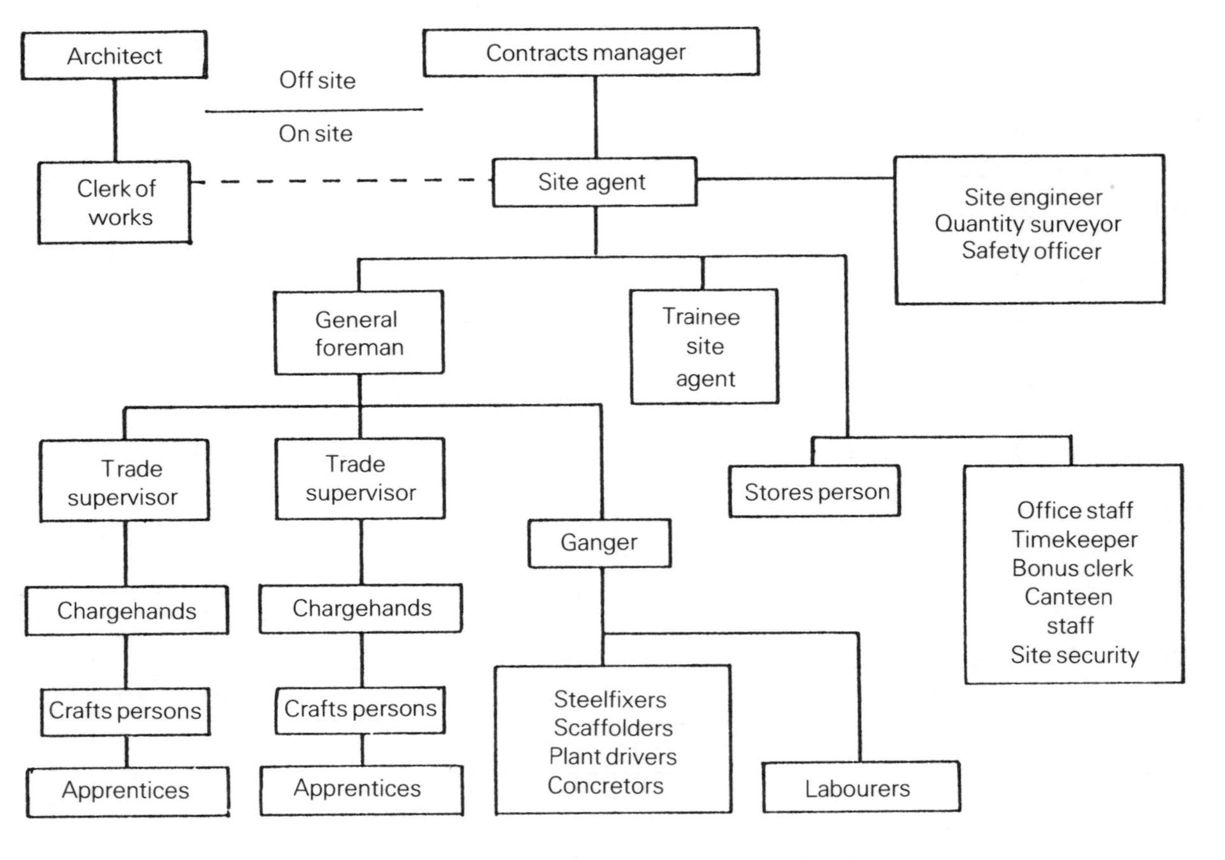

Organisation structure for a construction site

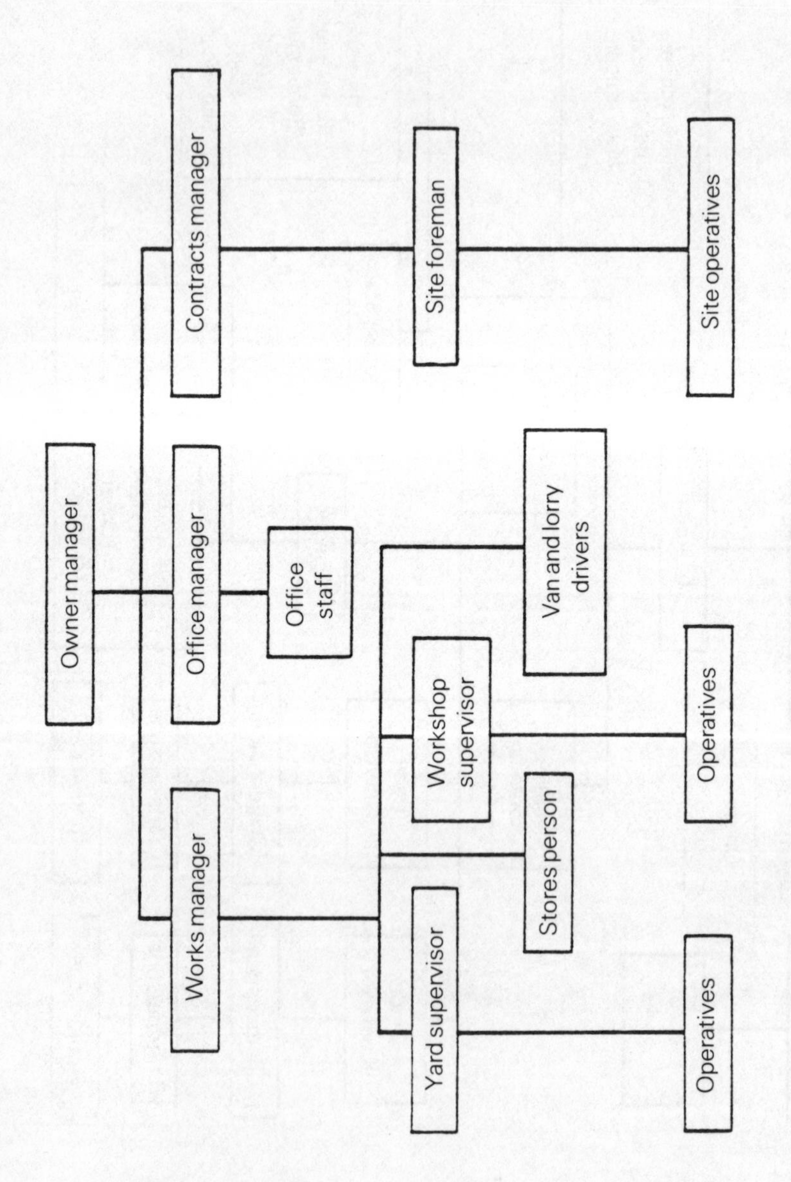

Organisation structure of a small firm

7 The People Involved in the Construction Process

THE MEMBERS OF THE BUILDING TEAM

There are many people involved directly or indirectly in the construction of buildings. The people involved can be placed into three groups:

(1) The designers and administrators.
(2) The constructors.
(3) The controllers and regulators.

The designers and administrators are employed by the client or by the client's architect, to create between them a design for the building which must fulfil as near as possible the needs of the client. They must also ensure that the building is constructed by the builders to the chosen design within the agreed financial limits.

The constructors are the building contractors. They are employed by the client to erect the building as designed with the minimum cost and the greatest possible speed.

The controllers and regulators are only involved with the building operations indirectly. Their job is to ensure that the relevant laws, regulations, restrictions and rules are complied with. Most of them are employed by central government and the local authorities.

THE MEMBERS OF THE BUILDING TEAM

The client.

THE MEMBERS OF THE DESIGN TEAM

Architect.
Consulting quantity surveyor.
Civil engineer.
Structural engineer.

THE MEMBERS OF THE CONSTRUCTION TEAM

The contractor.

Contracts manager.
Site agent or site manager.
Contractor's quantity surveyor.
Site engineer.
Surveyor.
Safety officer.
General foreman.
Trades supervisor.
Ganger.
Chargehands.
Craftsmen.
Semi-skilled operatives.
General operatives or labourers.

Sub-contractors.

There are also on site other personnel who play an important role supporting the construction team:

Storeman.
Wages clerk.
Bonus clerk.
Timekeeper.
Security staff.
Nurse.
Van and lorry drivers.

Off site in the contractor's office a number of personnel are involved in the day-to-day activities on site:

Estimator.
Buyer.

Client's representative on site

Clerk of works.

The regulators and controllers

Local authority technical officers.
Local authority building control officer.
Health and safety executive inspectors.

Other visitors to the site

Union officials.
Police.
Fire officer.

The client

The client is the person or organisation that commissions the building work and directly or indirectly employs everyone else.

Clients can vary from private individuals, small businesses, local authorities or pension funds to groups of organisations such as local authority consortia, small businesses etc.

Types of client	*Example*
Central government.	Ministry of Defence.
Local authorities.	Kent County Council.
Nationalised industries.	National Coal Board.
Public companies.	Wimpey PLC.
Partnerships.	Smith, Jones & Partners.
Sole traders.	Self employed persons.
Housing associations.	Cayster Housing Trust.
Private individuals.	Raymond Stacey.

The architect

The architect is employed by the client to act as his agent. He has a number of tasks that he must complete successfully during the design and construction stages before the client can occupy the building.

The architect's job is generally to:

- Design the type of building that the client has asked for, making sure that his design comes as close as possible to the required cost.
- Recommend a suitable builder to construct the building at the most competitive price available.
- Control and supervise the construction of the building.
- Ensure during the construction of the building that it complies with his drawings and instructions.
- Make sure that the cost of the work stays within the budget.
- Generally act for the client and when necessary advise him of the correct course of action to take.

Architects can work for themselves in private practice or be employed by another architect, a construction company, a local authority or a government department.

Until recently an architect could not advertise to obtain work, and had to rely on his reputation to obtain work.

Most architects are members of the Royal Institute of British Architects.

The consulting quantity surveyor

The quantity surveyor is chosen by the client, often on the advice of the architect. He is an important member of the design team. It is his job to advise the design team of the approximate cost of their various designs and then to monitor actual costs as the project progresses. The quantity surveyor is in effect the design team's accountant.

From the architect's design drawings the quantity surveyor measures all the quantities of the various materials necessary for the construction of the building. These quantities are written into a bill of quantities which is used by the contractor as a basis for his pricing of the work.

During the contract the quantity surveyor has to prepare valuations of the work done by the contractor so that regular payments can be calculated. If any alterations to the original design are to be made, he will also work out the difference in costs.

The structural engineer

The structural engineer is commissioned by the client to design and calculate the structure of the building. He has to work very closely with the architect to solve any structural problelms that may occur. The structural engineer is responsible for the structural stability of the building. He has to submit his design and calculations to the local authority for approval at the same time as the architect submits his plans.

The civil engineer

The civil engineer is an expert at stabilising soil, controlling water in the ground and reclaiming land. Many of his problems and tasks are similar to those of the structural engineer. Normally the civil engineer is engaged on the construction of projects such as roads, bridges, harbours, dams, canals, high rise buildings etc.

The services engineer

Modern buildings contain many specialist services such as plumbing, heating, drainage, refuse disposal, security systems, telecommmunications, and services for the movement of materials and transportation of people.

The services engineer has to work closely with the architect and structural engineer to find the best route for the services through the building, and to design the systems.

The clerk of works

On large and complicated sites the architect cannot always be available on site for consultation. The clerk of works is the client's representative on site and reports to the architect. He acts as the architect's eyes and ears on site, offering advice to the builder and liaising with the architect when problems crop up. If the architect thinks it is necessary, the clerk of works can issue instructions to the builder, on his behalf, altering the work.

The contractor

The contractor, as the name implies, contracts to undertake part or all of the construction work required by the client.

Main contractors undertake the major part of the construction work and organise the activites on site.

Sub-contractors are generally under contract to the main contractor to complete parts of the construction work that the main contractor is unable or unwilling to undertake.

The contracts manager

The contracts manager is employed by the contractor to supervise the running of a number of different contracts. He acts as a link between the contractor's office and the site agents under his control.

The site agent or site manager

The site agent has complete responsibility for all the activities that take place on site and the control of all site personnel.

His responsibilities include:

- Controlling site operations.
- Programming and planning work on site administration.
- Labour relations.
- Communications between site and the contracts manager or head office.
- Liaising with the architect.
- Dealing with official visitors to the site, e.g. union officials, fire officers, building control officers etc.

The site engineer

The site engineer sets out from the plans all the roads, drains, sewers and structures involved in the construction operations. During the construction of the work on site he will set out all the necessary levels and check that the work is vertical.

The contractor's quantity surveyor

The contractor's quantity surveyor works in co-operation with the site manager or site agent. It is his job to measure up the amount of work that the contractor has completed each month. From his measurements he calculates the value of the work done and checks that the monthly payments from the architect are correct. If the contractor has sub-contractors working on the site, the quantity surveyor will work out the amount that each one should receive as his monthly payment for the work he has completed. The quantity surveyor's calculations are also sent to the head office so that the contractor can assess his financial situation. It is very important for the contractor to be able to calculate whether the work is still within his budget.

The surveyor

The surveyor is employed on site to assist the quantity surveyor and to produce measurements of the work done for bonus payments to the men on site, payments for the sub-contractors, checking payments from the architect and working out the final accounts.

The safety officer

The safety officer is employed by the contractor to inspect and advise on the safe operation of the construction work. He can be employed on one large site or be responsible for a number of smaller sites. Some smaller contractors share a safety officer who will visit each of their sites in rotation.

The general foreman

The general foreman is responsible to the site agent for the smooth trouble-free progress of the construction work on site. It is his job to ensure that all the operatives engaged in the construction of the building are fully employed on productive work at all times. When a problem arises on site he will work closely with the site agent to achieve the most satisfactory solution. The general foreman's decisions may involve transferring some of the workforce to different parts of the site or other sites. He may also have to decide whether to increase or decrease the workforce by employing or dismissing operatives.

The trades supervisor

There are many different types of craftspersons on a site. The trades supervisor takes charge of a particular trade. He could be in charge of all the bricklayers for instance. He is responsible to the general foreman for keeping all the craftspersons in charge operating at maximum efficiency and productivity.

The ganger

The ganger acts in the same way as the trades supervisor but controls all the semi-skilled operatives and unskilled labour on the site. Their activities may range from laying concrete to general labouring.

The chargehand

On very large sites there may be a considerable number of operatives of a particular trade. It is often very difficult for the trades supervisor to control effectively the activities of every operative. Consequently, the craftspersons are divided into gangs or teams under the control of a chargehand. The chargehand controls his group of craftspersons under the supervision of the trades supervisor.

The craftsperson

The craftspersons are operatives who possess the specialist skills needed to perform skilled tasks and work with specific materials.

Examples

bricklayer	bricks	build a wall
carpenter	timber	hang a door
painter	paint	decorate a door
plumber	copper pipe	connect a tap

Semi-skilled operatives

Semi-skilled operatives undertake activities that do not require the high levels of skill or training that a craftsperson possesses.

Examples

concretor	laying concrete
tiler	laying floor and wall tiles

General operatives or labourers

There are a lot of activities on site that need little or no skills to be completed successfully. These activities are often undertaken by general operatives or labourers.

Examples

- Digging trenches.
- Moving materials.
- Assisting semi-skilled operatives.
- Assisting craftspersons.

The stores person

Very large sites have enormous quantities of materials delivered. These materials need to be issued to the people on site when needed. Care is also needed in storing and handling some materials to reduce losses, damage and costs.

The stores person has to make sure that the correct quantities of materials are delivered on time as ordered by the buyer. An adequate supply of materials will allow the construction work to progress smoothly and efficiently without material shortages or overstocking.

The stores person's job involves:

- Ordering materials.
- Checking deliveries.
- Recording the issue of stock.
- Keeping enough materials in stock to satisfy demand.
- Keeping materials stored in good condition and secure from theft.

The timekeeper, wages clerk and bonus clerk

On large sites these jobs are done by different people but small sites may have only one person to undertake all the tasks.

The jobs involve:

- Recording the arrival and departure times of the employees on site.
- Calculating bonuses from timesheets.
- Preparing wages from the timesheets, bonus sheets, tax and insurance tables.
- Collecting money from the bank and making up wage packets.
- Giving out the employees' wages.

The buyer

To enable the contractor to purchase the correct quantities of materials at the lowest price and have them delivered on schedule, the contractor employs a buyer in his office.

It is the buyer's job to:

- Obtain quotations.
- Find out delivery dates.
- Order materials.
- Set delivery dates for orders.
- Ensure that ordered materials arrive on time in the correct quantities.

The estimator

The estimator's job is to estimate the total cost of the building work from the contract drawings and the bill of quantities. To find the cost of each item the estimator has to obtain material prices from the buyer and calculate the time it will take to do the work. The price he arrives at is the amount it will cost the contractor to complete the work. When profit is added to this price the contractor will have his tender price. The tender price is the price at which he will offer to undertake the work.

Local authority technical officers

Highway engineers.
Town and country planning officers.
Public health inspectors.

The local authority employs a variety of technical officers to regulate and control changes in the built environment. They ensure that work in their area of responsibility is carried out to the correct standard, They often make spot checks as well as regular inspections.

The building control officer

The building control officer examines the plans of all proposed building work. He makes sure that the proposed work complies with the building regulations before he will allow it to proceed. The building control officer, or building inspector as he is sometimes known, will regularly inspect approved work as it proceeds and on completion. If he finds the work is not complying with the building regulations, he has the power to enforce its alteration until it reaches the approved standard.

The health and safety inspector

The Health and Safety Executive has a number of branches controlling different sections of industrial activity. The Health and Safety Executive Inspectorate (or what used to be called the Factory Inspectorate) ensures that regulations covering safety, health and welfare are complied with in factories and on sites.

The health and safety inspector has the power to enter any workplace and has legal powers to enforce the regulations if necessary.

Union officials

Full time union officials have the right under the Working Rule Agreement to enter sites to inspect union cards. They can also discuss and negotiate any grievance or dispute that may arise with the different parties.

8 Design and Construction

THE BUILDING PROJECT FROM DESIGN TO COMPLETION

These flowcharts detail the activities of the architect and builder during the design, construction and completion of a building. As no two building projects are exactly the same, these flowcharts serve only as a guide to the possible sequence of activities. Depending on the size and complexity of the work, changes could occur in the operations and their sequence.

THE PROJECT FROM DESIGN TO COMPLETION

The architect's activities during the design and construction of a building

1

Initial interview

Before the client employs the architect and the architect commences work on the project they have a preliminary meeting. At this meeting they discuss the client's requirements. After this meeting the architect will be able to formulate some ideas for the design of the proposed building.

2

Site visit

After the initial discussion with the client the architect visits the site to obtain a general impression of the area around the site and to look at the site itself.

The builder's activities during the design and construction of a building

Architect's activities | **Builder's activities**

3

Briefing the architect

When the client has decided to appoint the architect and the architect accepts the work, the design brief is prepared during discussions and meetings. The brief describes the details of the required building and any other information related to it.
For example:
building type,
floor area,
number and type of rooms,
cost,
has outline planning consent been obtained,
who owns the land,
who else is involved in the project,
when will the site be available,
by what date is the building to be completed.

4

Site inspection

The site inspection is a detailed examination of the site. It provides all the relevant information about the site to enable the architect to design the building properly.
The inspection will give information on the:
site boundaries,
adjoining buildings,
features of the site,
direction the site faces,
surrounding landscape,
access roads,
footpaths to and across the site,
access to the services e.g. gas, water, electricity,
nature of the soil,
streams and rivers on or near the site,
local materials,

5

Local authority talks

Before becoming involved in more detailed work on the project the architect should at this stage discuss with the local authority his client's requirements and the proposed design. The local authority

Architect's activities

Builder's activities

will then advise the architect on the possibility of planning permission being granted.

6

Outline planning permission

Before any work can be carried out on the design of the new building the architect should obtain outline planning approval for the project from the local authority.

It would be a pointless and very expensive exercise to design in great detail a tall office block, for example, when the local authority will only allow the construction of a small block of flats on the site.

7

Site survey

A more detailed survey of the site is now needed. The site survey covers in greater detail all the areas in the site investigation as well as the following:

the geology of the area,
the nature of the sub-soil,
mine workings in the area,
the water table level,
the possibility of flooding,
site dimensions,
ownership of fences,
location of all services,
details of adjoining property and its owners.

8

The design team

The specialist members of the design team are now appointed e.g. quantity surveyor, services engineer, civil engineer etc.

9

Design Report

The architect now has:
(a) an idea of what his client wants his

Architect's activities | **Builder's activities**

building to look like,
(b) the information provided by the brief
(c) advice from specialists,
(d) an impression of the site and its surroundings.

From these items the architect can now produce a report containing:
(a) the proposed design including sketches of the building,
(b) a model of the proposed building if required,
(c) an estimate of the cost,
(d) any other recommendations.

10

Client's approval

The architect should now obtain the client's approval of the design report.

11

Basic plans

The architect produces basic design drawings of the building for the design team.

12

Design team's operations

The architect provides the members of the design team with:

information about the building,
the basic design drawings,
a programme of the work to be done.

From this information the design team will produce the construction drawings and schedules, in close collaboration.

The architect must ensure that the design team work together and are kept informed of each other's progress.

13

Construction drawings

The detailed construction drawings are produced. As some alterations will occur they must be kept up to date throughout the detailed design stage.

Architect's activities

Builder's activities

14

Client's approval

The architect will obtain the client's approval of the construction drawings as they are produced.

15

Local authority consent

Full planning permission and building regulation approval must now be obtained from the local authority.

16

Schedules

The schedules are produced. These will clearly give information on items such as doors, windows, sanitary and electrical fittings that cannot be clearly detailed on the drawings.

17

Client's approval

The architect should now obtain the client's approval of the proposed nominated suppliers and subcontractors.

18

Bill of quantities

The quantity surveyor produces the bill of quantities from the construction drawings.

19

Client's approval

The architect should now obtain the client's approval of the bill of quantities.

Architect's activities

20
Estimated cost

From the bill of quantities an estimate of the total cost of the work is made.

21
Client's approval

The architect should now obtain the client's approval of the estimated cost of the building.

22
Clerk of works

A clerk of works may be appointed to act as the architect's agent on site.

23
Tender list

A list of main contractors, previously approved by the client, is prepared. They are to be invited to tender for the work.

24
Tender documents

The tender documents are prepared for issue to all the contractors tendering for the work.
They are the:
- invitation to tender,
- bill of quantities,
- specification,
- drawings,
- schedules,
- tender form.

25
Invitations to tender

The architect invites the contractors on the list to submit tenders for the work. A set of tender documents is sent with this invitation.

The builder's activities during the design and construction of a building

The builder's activities are divided into four main sections:
- pre-tender work,
- pre-contract work,
- the construction process,
- post- construction work.

PRE-TENDER WORK

1
Invitation to tender

The architect invites the builder to tender for the contract.

Architect's activities **Builder's activities**

2

Reply to invitation to tender

The builder inspects the tender documents. If he is able to undertake the work, he accepts the invitation.

He may refuse the invitation because the work does not fit in with his other contracts, the contract is too large for him to undertake, or the contract does not involve the type of work his firm normally undertakes.

3

Collect Information

After accepting the invitation to tender the builder assembles as much information as possible. The information enables a pre-tender report to be produced.

The information for the report will include details of:

the contract documents,
the estimated contract time,
the site and its surrounding area,
the availability, cost and delivery dates of materials,
the availability of labour,
possible construction methods.

4

Pre-tender report

On the information provided by the pre-tender report the builder's team can plan the work. They will determine the best method of undertaking the work. The method selected will establish how the lowest tender price can be obtained.

5

Method statement

The method statement specifies the way in which the work is to be done. It details the:

operations,
total quantities,
methods,

Architect's activities | **Builder's activities**

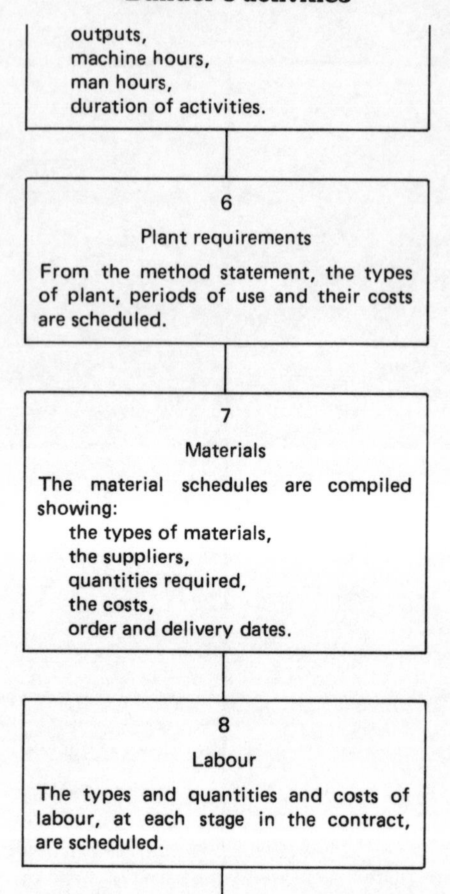

9

Site organisation

The best method of organising the layout of the site is established.

10

Pre-tender programme

On the basis of the available information, a time-based programme of the construction work is produced. The times of the major operations are used to detail the programme. The total construction time must not exceed the contract period.

Architect's activities

26

The tenders

The tenders that have been received on time are all opened by the architect at the appointed time. The quantity surveyor, client and contractors may be present for the opening.

27

Lowest tender

The lowest tender is usually selected. The second lowest tender is held in reserve in case any problems occur with the lowest tender.

Builder's activities

11

Calculate unit rates

The rates for labour, plant and materials are calculated. They are based on the plant, labour and materials information already scheduled.

12

Price bill of quantities

The bill of quantities is priced using the calculated unit rates.

13

Prepare tender

On the basis of the priced bill of quantities a tender price is prepared.

14

Submit tender

The sealed tender for the contract is submitted to the architect.

15

Attend tender opening

The builder, if he wishes, can attend the opening of the tenders at the architect's office.

Architect's activities

28

Tender price

The quantity surveyor checks the priced bill of quantities of the lowest tender to ensure that its prices and conditions are reasonable.

29

Client's approval

The architect should now obtain the client's approval of the lowest tender and its price.

30

Contract

The contract is made with the building contractor to undertake the work at the tendered price.

Builder's activities

16

Accept contract

If the builder's tender is acceptable he will be invited to enter into a contract for the work.

17

The contract

The builder contracts to undertake the work at the accepted price.

PRE-CONTRACT WORK

18

Review pre-tender planning

After being awarded the contract more detailed information is needed by the builder. The initial information is brought up to date, revised and expanded.

19

Construction programme

A detailed programme of the construction work is produced. The major operations in the pre-tender plan are broken down and programmed in greater detail.

Architect's activities	**Builder's activities**

20

Activity programmes

Detailed programmes are produced based on the master programme. They show:
- material order and delivery dates,
- plant requirements,
- labour quantities and types.

21

Produce site layout

The layout and organisation of the site is detailed. The layout will show the positions of:
- site offices,
- site accommodation,
- toilets,
- stores,
- storage areas,
- plant,
- plant compounds,
- temporary roads,
- hoardings and entrances.

22

Orders

Orders are placed for materials and plant, stating the quantities required and their delivery dates.

23

Acquire labour force

Prior to commencing work on site the contractor must obtain a suitable labour force. The transfers of existing employees are scheduled and any new employees obtained.

24

Services

Water, electricity, and telephone connections to the site are arranged just prior to commencing work on site.

Architect's activities

31

Management during the construction period

Throughout the construction of the building the architect is involved in the management of the project. His activities ensure that the work progresses on time and to plan.

Site meetings

Regular site meetings are held between the architect, quantity surveyor, clerk of works, contractor and subcontractors to discuss:

- progress of the work,
- modifications ot the work, and the prevention of delays.

Site reports

The architect receives regular site reports from the contractor and the clerk of works. These reports keep the architect fully informed of the:

- daily weather conditions,
- labour force on site,
- materials and plant delivered,
- shortages of labour, plant and materials,
- delays and stoppages.

Progress reports

The architect sends regular reports to the client on the progress of the work.

Financial reports

The architect must keep a close check on the cost of the work. Regular financial reports, prepared by the quantity surveyor are sent to the client.

Builder's activities

THE CONSTRUCTION PROCESS

25

Management during the construction period

During site operations the contractor is responsible for supervising and administering all the work on site.

26

Review and revise the programme of work

Throughout the period of the construction process the contractor must continually review and revise the progress of the work.

Review method statement

The contractor reviews the information contained in the method statement throughout the contract. The progress of the work is monitored and any alterations to times and quantities noted.

Review construction programme

Amendments to the project, bad weather and delays can occur. Revision of the construction programme will allow scheduled operations, order dates and delivery dates to be easily amended.

Check progress

Progress is continually monitored as the work proceeds. Any delays to the programmed work are easily detected. If necessary suitable remedies are put into effect.

Architect's activities

Variations and modifications

During the building work problems or changes will alter the original design. Extensions, variations or modifications to the original design will require new and altered drawings and schedules. The architect will have to produce the new information and control any changes that may occur in the cost.

Interim certificates

The architect issues interim certificates, prepared by the quantity surveyor, to the contractor each month. The certificate states the value of the work done during the month by the contractor. When the client receives an interim certificate he will pay the amount stated on it to the contractor. A part of the money due will be kept back in a retention fund until the end of the contract.

Maintenance file

During the construction of the building the architect collects information for the maintenance file. The client will need this information to maintain his building on completion.

32

Inspection

When the building is completed, the Architect, with the contractor and client, has a final inspection of the building to ensure it is up to the required standard.

Builder's activities

Site diary

During the construction of the building the contractor records daily the:

- weather conditions,
- labour force on site,
- materials and plant delivered,
- shortages of labour, plant and materials,
- delays and stoppages.

Site reports

The contractor sends regular reports to the architect. The reports inform the architect on:

- the progress of the work,
- weather conditions,
- labour force on site,
- materials and plant delivered,
- shortages of labour, plant and materials,
- delays and stoppages.

Site meetings

Throughout the duration of the contract regular meetings are held between the contractor, architect, quantity surveyor and clerk of the works to discuss:

- progress of the work,
- modifications to the work and the prevention of delays.

The construction activities

The construction of the building proceeds under the control of the contractor according to his programme of work.

The major construction activities are:

- set out the site,
- excavate the foundations,
- build the substructure in the excavation,
- build the superstructure,
- install services,
- install finishes.

Architect's activities

33

Completion

The building is completed when the architect is satisfied with the work and the builder has left the site.

34

Completion payment

On completion the architect will tell the quantity surveyor to pay the builder 50% of the money in the retention fund.

35

Occupation of the building

The architect tells the client he can now occupy the building.

36

Final account

The architect instructs the quantity surveyor to prepare the final accounts.

37

Defects period

During the first six months of occupying the building the client notes any defects that occur. The architect will then instruct the contractor to remedy these defects if they are due to his bad workmanship.

38

Final certificate

The architect issues the contractor with the final certificate. The contractor presents the certificate to the client who will complete the contract by paying him all the money due.

Builder's activities

POST-CONSTRUCTION WORK

27

Completion and handover

When the building is completed the architect inspects the work. If he is satisfied the work is complete, the contractor hands over the building to the client.

28

Defects liability

During the defects liability period the contractor corrects any defects in the building due to bad workmanship.

29

Settlement of final account

When the building has been completed the final account is settled between the contractor and the client.

Architect's activities

39

Final fees

The fees, unless included in the interim certificates, for the quantity surveyor and other consultants who formed the design team are sent to the client.

40

Settlement of accounts

The architect checks that all the bills have been paid.

41

Details of the building

The architect's final job is to file all the records of the building for future reference.

Builder's activities

30

Analysis of job

When the contract is completed the builder conducts an examination of the project.

The estimated costs are checked against actual costs. The times used to programme the work are examined. The causes and results of delays are listed.

The information obtained is used to revise the builder's tendering and programming methods.

9 The Construction of a Building

The construction of a building, no matter how small, needs to be done efficiently, to schedule and within the tender price. To achieve this the builder must carefully plan and control his activities by using a number of planning tools. These tools determine the precise sequence of construction activities, the methods employed, and the time allowed for each activity. They are usually produced in tabular or chart form, which are easy to read and understand. They are used in the builder's office and on site in the form of:

Statements.
Programmes.
Timetables.
Schedules.

Method statement

A method statement describes the methods that will be employed to complete an activity. It is usually in tabular form. The table consists of a number of columns covering the following areas:

Operation.
Quantity and description.
Method.
Output.
Man or machine hours.
Duration of operation.
Remarks.

Under each column the descriptions of the individual operations in the construction sequence are listed.

Operation	Quantity	Description	Method	Output	Man hours/ machine hours	Duration of operation	Remarks
Excavate trenches for strip foundation.	24 cubic metres.	Excavation of foundation trench to minimum of 1.5 metres deep in firm clay on dry site.	Excavator with a 1 cubic metre bucket. Lorries (2) to remove spoil from site.	10 cubic metres/ hour. 5 cubic metres/lorry load.	Excavator driver 12 hours. Lorry drivers 20 hours Excavator 10 hours. Lorries 20 hours.	10 hours.	Journey time to tip for lorries is 15 minutes each way.
Pouring strip foundations.	5 cubic metres.	Pour concrete strip foundations to a minimum depth of 300mm in bottom of trenches.	Unload from mixer lorry directly into the foundation trenches.	Mixer lorry 5 cubic metres.	Labourers 3 hours	1 hour.	

A section of a simple method of statement showing the excavation and pouring of a concrete strip.

Programme of work

More information can be obtained by examining the data already produced. An examination of the method statement will provide details of plant requirements. The programme of work will show when each item of plant is required on site and for what length of time. A bar chart constructed from this information will give a timetable showing plant requirements. Information can be added to show when the plant needs to be ordered and delivered on site.

The same process can be used for ordering and delivering materials. A similar process can be used to construct a vertical bar chart called a histogram showing the quantity and type of operative on site.

The following examples of bar charts show the sequence of construction for part of a detached house.

Plant timetable

The information about the plant required can be obtained from the timetable. It is usually in the form of a bar chart and shows the periods when each item of plant is on site. It can also show order dates, delivery dates, idle time etc.

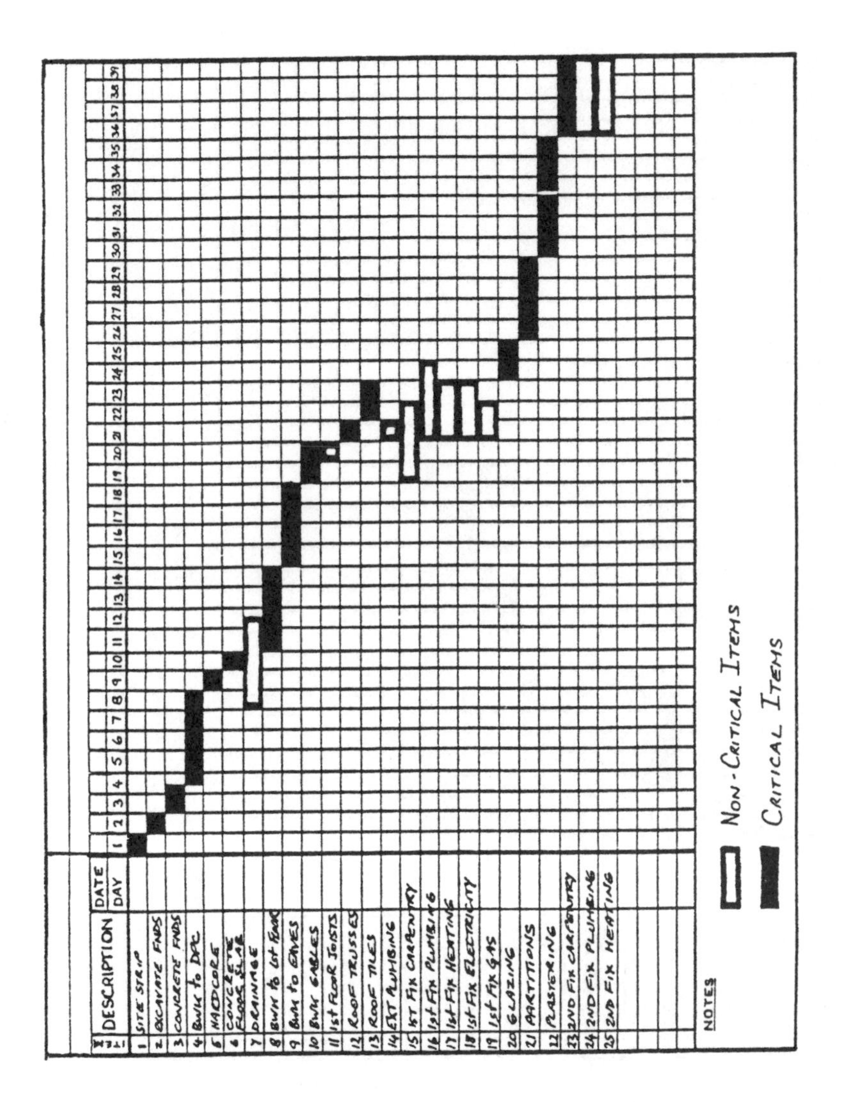

Bar chart – programme of construction activities.

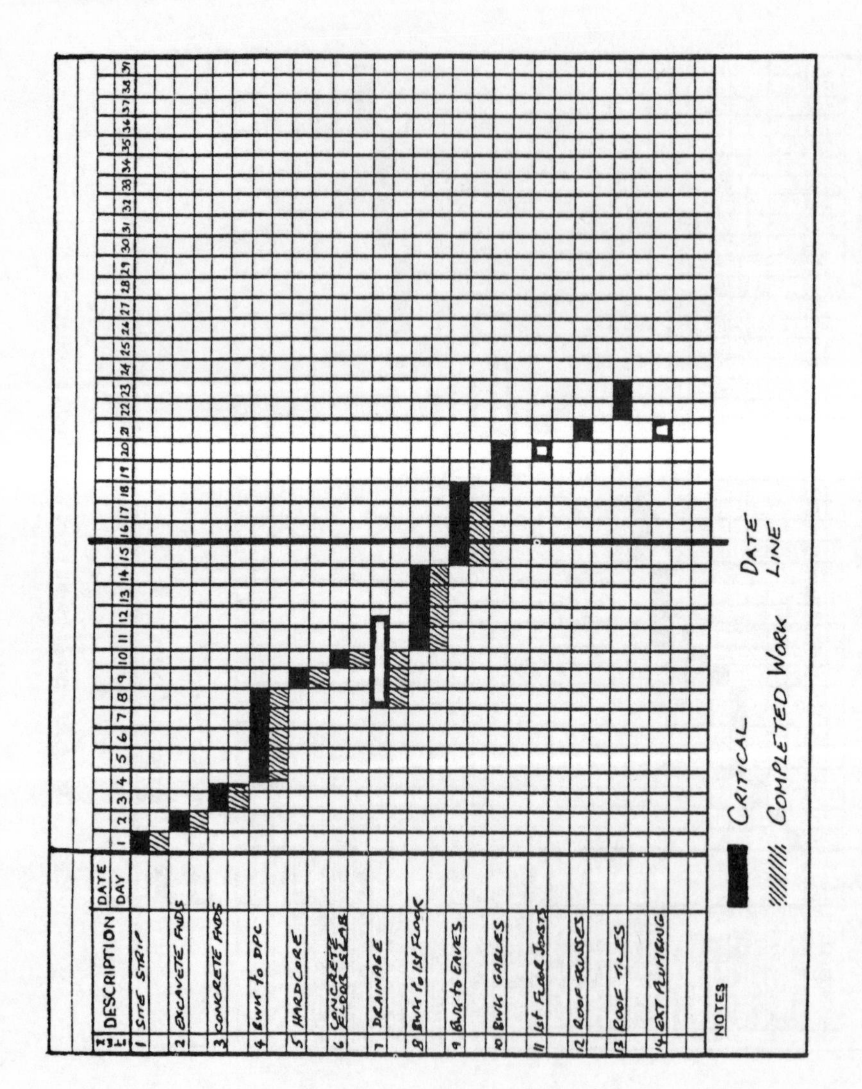

Bar chart – programme of construction activities showing progress to date.

CASE STUDY

The garage

Mr Green has recently bought a detached house with a large garden. Unfortunately, the house does not have a garage and Mr Green decided to have one built. After listing out what he wanted building, he told his architect to produce a suitable design based on his list.

Mr Green's list

A double garage with a work area at the rear.

An up-and-over door giving access to the garage from the road.

A glazed door and large window at the rear of the garage giving access from the patio.

The drive and paths to the road should be of tarmacadam.

The patio should face south and be secluded.

The paving on the patio should be of stone and match the colour of the brickwork.

The garage and patio should have sufficient lighting to enable them to be used at night.

Water taps at the front of the garage and on the edge of the patio by the garden.

Mr Green also instructed the architect to place the following conditions into the contract for the work:

Construction time 16 days maximum.

The work must be done in May while Mr Green is on holiday.

The builder will be allowed to set up the site on the Friday afternoon.

The work must be completed by the Saturday evening when Mr Green returns.

The architect

When the architect had produced a suitable design, planning permission was obtained. Detailed drawings, a 'mini' bill of quantities, and a specification were then prepared. These were sent out to a selected number of local builders for tender. ABC Construction's price for the work was accepted and a contract was made with them to do the work.

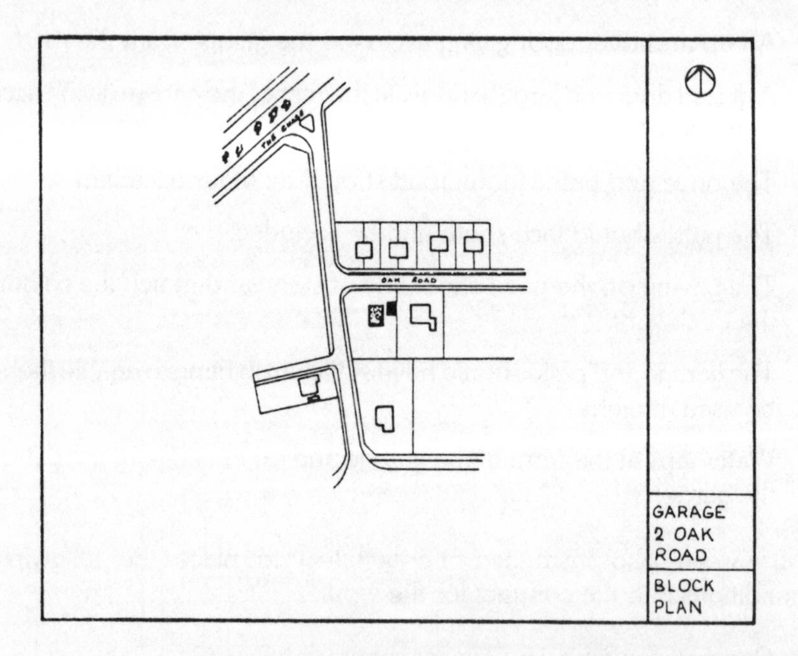

Block plan (typically drawn to scale 1:1250).

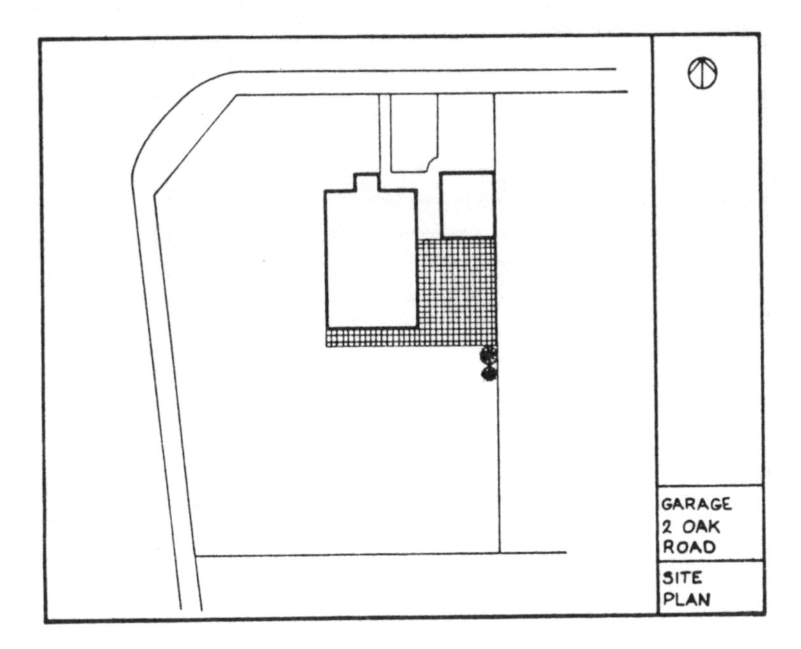

Site plan (typically drawn to scale 1:500) showing the proposed layout of the site.

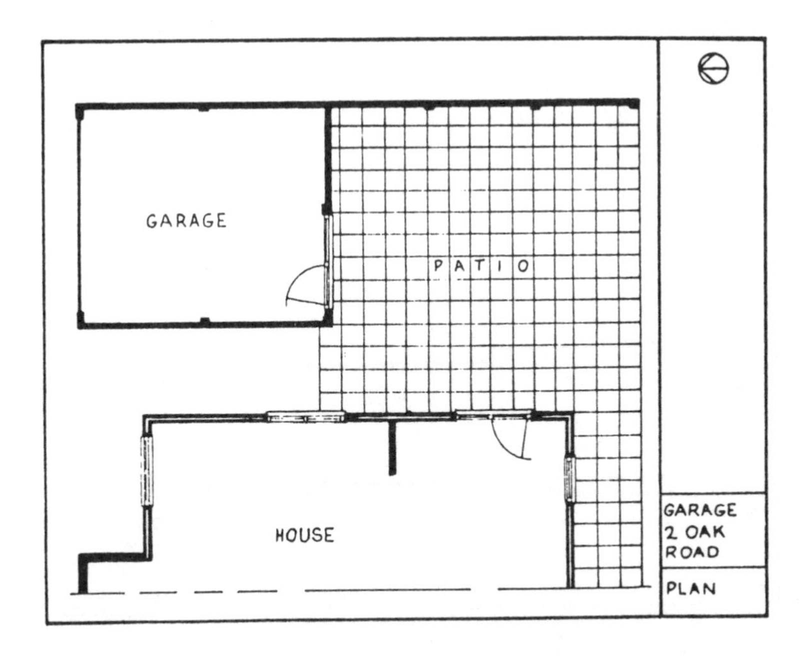

Location plan (typically drawn to scale 1:200) showing details of the garage.

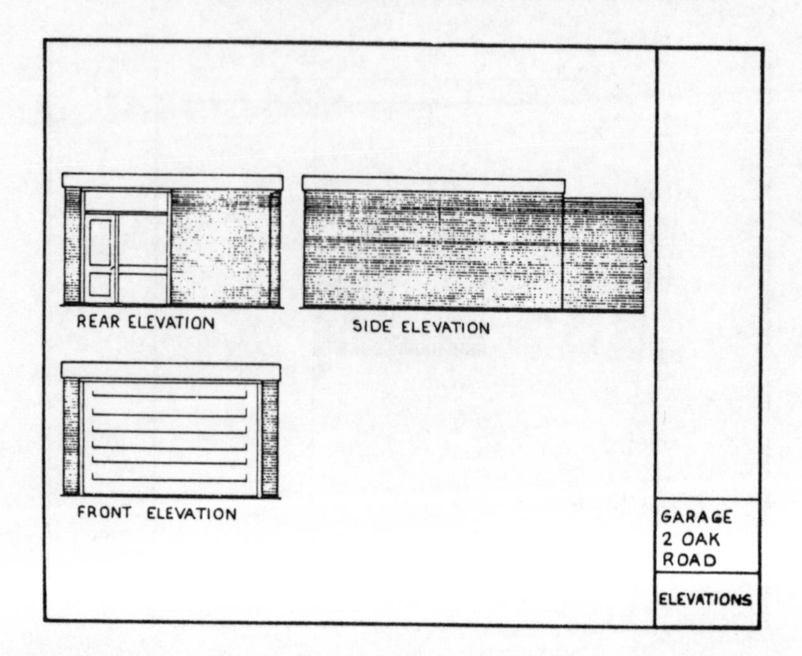

Plan (typically drawn to scale 1:200) showing elevations of the garage.

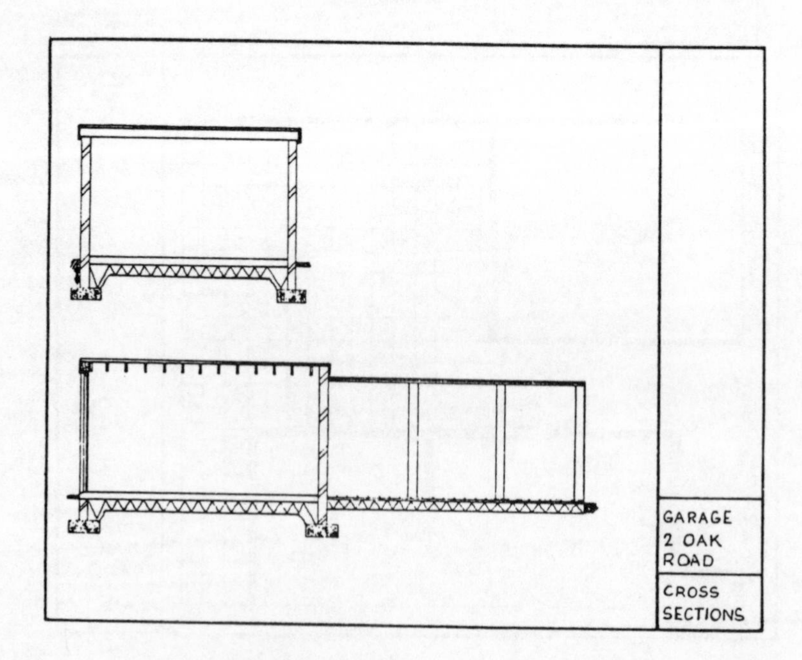

ABC Construction

During the tender stage a programme for the work and a method statement had been produced. The time limit for the work on site had caused problems and difficult decisions had to be made. The choice was between making the maximum use of expensive plant, or employing large numbers of men to do the work. The use of plant was considered more productive and cost effective. Employing large numbers of men on site could result in delays and inefficiency as they could obstruct each other and prevent efficient working.

It was also decided when producing the method statement to work Saturdays and Sundays. These extra five days would increase the time on site by 50 per cent.

On obtaining the contract, a review of the pre-tender plans was made and detailed plans of the construction work produced. The review produces a detailed method statement, programme of work, and plant timetable.

PROJECT GARAGE & PATIO PROGRAMME of WORK

ITEM	DESCRIPTION
1	Set up site
2	Clear & Level Site
3	Excavate Foundations
4	Concrete Foundations
5	Services below Ground
6	Brickwork to Finished Floor Level
7	Hardcore Slab
8	Concrete Slab
9	Brickwork to Eaves
10	Flat Roof
11	Hardcore Patio
12	Brickwork Patio Wall
13	Pave Patio
14	Garage Doors
15	Painting
16	Services
17	Clear Site

DATE
DAY 1 2 3 4 5 6 7 8 9 10 11 12 13 14 15

CRITICAL
NON-CRITICAL

Bar chart – programme of work for the garage.

ITEM	PROJECT GARAGE & PATIO PLANT TIMETABLE																	
	DESCRIPTION	DATE / DAY		1	2	3	4	5	6	7	8	9	10	11	12	13	14	15
1	Site																	
2	Excavator																	
3	Dumper																	
4	Cement Mixer																	
5																		
6																		
7																		
8																		
9																		
10																		
11																		
12																		
13																		
14																		
15																		
16																		
17																		

Bar chart – plant programme.

Method statement

Operation	Quantity and description	Method	Output	Total man hours	Machine hours	Duration of operation (days)	Remarks
Clear site, excavate to reduced levels and cart away excess soil.	40 m^3 max 450 mm deep firm clay dry site.	Excavator with ½ m bucket and lorry to cart away soil.	Excavator to dig and load 5 m^3/hr into 5 m^3 lorry. Trip to tip 30 mins	24	12	1½	
Excavate trenches for strip foundations.	24 m^3 max 1.5 m deep firm clay dry site.	Excavator with 600 mm bucket.	Excavator to dig and heap 1½ m^3/hr.	16	16	2	
Concrete strip foundations.	6.0 m^3 concrete in 600 x 225 mm foundations.	Ready mix concrete in 5 m loads. Concrete to be placed directly into trench. Two labourers.	¾ m^3/man/hr including tamping.	8	–	½	
Lay service pipes and cables.		Lay pipes and cables in trenches.		12	–	1½	
Brickwork to finished floor level.	7 N/mm sand lime bricks in 1:3 sand/cement mortar. 18 m^3 of/brick wall.	One bricklayer and one labourer. Mortar mixed on site in mixer.	1½ m^2/bricklayer/hr	12	4	1½	
Lay hardcore for garage floor.	16 m^3 hardcore for garage.	Dumper. Two labourers and consolidating.	½ m^3/labourer/hr	16	10	1	
Concrete garage floor slab.	8 m^3 concrete floor slab.	Ready mix concrete in 5 m loads. Concrete to be placed direct to base. Two labourers.	1 m^3/labourer/hr	8	–	½	
Brickwork to eaves.	21 N/mm wire cut facing bricks in 1:3 sand/cement mortar. 42 m^2 of/brick wall.	One bricklayer and one labourer. Mortar mixed on site.	1.5 m/ bricklayer/hr	56	20	3½	

10 Construction Safety

THE CONSTRUCTION INDUSTRY'S SAFETY RECORD

The type of work undertaken by the construction industry is often difficult and hazardous. Most hazards cannot be eliminated by just providing physical protection. The type of work and conditions are different on each site. Consequently, early planning to establish safe working methods must be undertaken. Managerial staff and operatives must work together to ensure that safe working is the prime consideration on site.

When the record for the whole of manufacturing industry is compared with that of the construction industry the facts are clear. The construction industry has a fatal accident rate five times greater than that of manufacturing industry.

Any industry with a high rate of fatal accidents and serious injuries must be considered to be inefficient. The financial losses due to lost production and man hours are high, losses an industry can ill afford to bear. Society also has to pay a price, as social and medical care are expensive.

Accident statistics

Accidents which cause death, major injury or involve more than three days absence from work have to be reported to the Health and Safety Executive. The following tables show the annual totals for reported accidents issued by the Health and Safety Executive. These tables compare the accident rates for manufacturing industry with those of the construction industry.

Fatal accidents	Manufacturing industry		Construction industry	
Year	Deaths	Per 10 000 employees	Deaths	Per 10 000 employees
1983	118	2.2	118	11.6
1984	142	2.7	100	9.8
1985	124	2.4	104	10.5
1986–87	109	2.1	99	10.2
1987–88	98	1.9	102	10.3
1988–89	92	1.8	99	9.7
Average	113.8	2.18	105.5	10.4

Major accidents	Manufacturing industry		Construction industry	
Year	Total	Per 10 000 employees	Total	Per 10 000 employees
1983	4308	79.6	2178	213.2
1984	4758	89.6	2288	225.2
1985	4866	92.3	2239	225.8
1986–87	7231	141.4	2570	263.5
1987–88	7065	140.1	2633	265.2
1988–89	7091	137.5	2701	265.3
Average	5176	113.5	2435	242.7

Accidents involving more than three days off work	Manufacturing Industry		Construction industry	
Year	Total	Per 10 000 employees	Total	Per 10 000 employees
1986–87	53767	1051	16260	1667
1987–88	52538	1042	16418	1654
1988–89	54811	1063	15963	1568
Average	53705	1052	16214	1630

In the five years from 1984 to 1989, 504 construction workers lost their lives and 124 000 were seriously injured – an unacceptable level of accidents that makes the construction industry one of the most dangerous, especially for those who are self employed. The construction industry accounts for nearly 70% of all injuries to self employed people.

Workers killed	Seriously injured	Rate
101	24 000	Yearly
2	477	Weekly
0.3	68	Daily
	8.5	Hourly

Safety legislation

In the early part of the nineteenth century working men, women and even children laboured in terrible working conditions which caused many occupational diseases. The high rates of disease, industrial injury and death caused a public scandal. Lord Shaftesbury's Factory Act of 1833 created legal controls designed to remove unsafe practices and improve working conditions.

The use of new machinery, new processes and hazardous operations are not adequately covered by the Factory Acts. Extra regulations are made to cover special areas of work without amending the main piece of legislation, e.g. the Abrasive Wheel Regulations 1970.

In 1972 the Robens report was published. It examined all aspects of health and safety at work. The findings of this report resulted in the Health and Safety at Work Act 1974.

The Factory Act 1961 and the Health and Safety at Work Act 1974 are the main pieces of legislation covering health and safety at work. The introduction of this legislation has improved the standards of industrial health and safety. Working conditions and the workplace are now safer and healthier.

Health and Safety at Work Act 1974

The Health and Safety at Work Act 1974 and its subsequent legislation were designed to encourage the employer and employee to control dangers to health and safety themselves. They should identify potential dangers and rectify them, rather than comply with the minimum legal requirements.

The Health and Safety Executive

The Health and Safety Executive is a government agency that enforces the Health and Safety at Work Act.

The Health and Safety Executive's inspectors have considerable powers. They can:

- Enter a firm's premises or site to make examinations and investigations.
- Be escorted by a police officer if obstructed in their work.
- Have premises left undisturbed to enable investigations to take place.
- Obtain measurements, samples and photographs.
- Obtain any relevant documents, registers, records etc.
- Enforce any improvements in safety.
- Issue prohibition notices to prevent unsafe working.
- Take signed statements from individuals.
- Take criminal proceedings against employers and employees.

Site safety

A SAFE SITE

A site can be considered safe when people are able to go about their normal activities without undue risk.

ACCIDENTS

An accident is when an uncontrollable chain of events causes injury or damage.

ACCIDENT PREVENTION

Accident prevention is achieved by exercising proper control in the workplace. Safe working is a result of employing correct methods and procedures.

CAUSES OF ACCIDENTS

Accidents can be caused by:

- A lack of discipline and training.
- Poor communications.
- Carelessness and ignorance.
- Distractions and a lack of concentration.
- Failure to wear or use the correct safety equipment.
- Incorrect handling and lifting of heavy objects.
- Misuse of power tools, plant, machinery and transport.
- Untidy and badly organised sites.

THE COST OF AN ACCIDENT

Accidents are paid for in human suffering and in financial loss by the employee and employer. Even a minor accident can be expensive:

- Wages have to be paid to the operative even though he has not earned them.
- Wages have to be paid to other operatives whose work is halted, even though they have not earned them.
- Wages have to be paid to staff investigating and reporting the accident.
- Wages have to be paid for a replacement to do the work of the injured operative.
- Production is halted during and after the accident.
- The planned sequence of operations is disrupted.
- Damaged plant, materials and equipment must be paid for.
- Plant and transport taken out of service must be replaced.
- Administrative expenses must be paid for by the employer.
- After the accident productivity will initially be lower.

Cutting costs in the short term, at the expense of safety, may be an expensive exercise in the long term.

HAZARDS ON SITE

Supervisors and operatives should be aware of the possible dangers and hazards on site. If the workforce is safety conscious hazards will be reduced and site safety improved.

Types of hazard

Falls.
Falling objects.
Transport.
Electricity.
Machinery.
Fire and explosion.

Causes of accidents

The following pie chart shows the proportion of fatal accidents for each hazard. It can be seen that falls account for 24%, almost 1 in 4 of all fatal accidents.

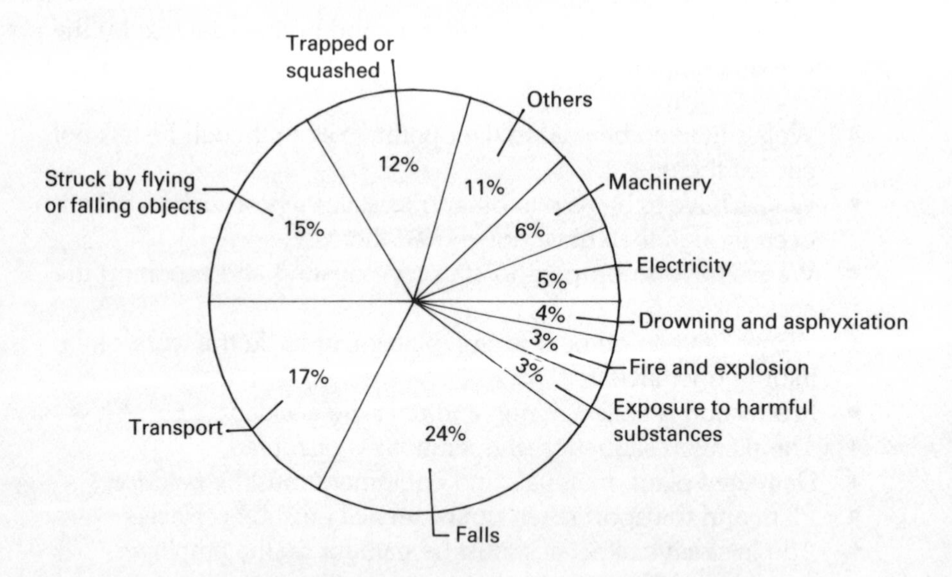

PERSONAL PROTECTION

If the workplace was perfectly safe, safety equipment would not be needed. Unfortunately, a perfectly safe workplace is difficult to achieve and maintain. Therefore, to improve the standards of personal protection, employers have to provide by law

- Eye protectors or eye shields for specific operations.
- Respirators when dangerous fumes and dusts are present and adequate ventilation is not practicable.

- Protective clothing against asbestos products when adequate ventilation is not present.
- Ear protectors when noise levels are unacceptable and cannot be reduced.
- Protective clothing for employees who have to work out in inclement weather.

The Working Rule Agreement also recommends that:

- Safety helmets should be worn where head injury is a risk.
- Consequently, they should be worn at all times in the construction industry.
- Personal safety can also be improved by wearing safety footwear and gloves.

Personal safety protection equipment.

11 Legislation

THE CONTROL OF BUILDING BY LEGISLATION

During the Middle Ages the population grew steadily. The standard of dwellings and public health was poor. Consequently, disease and plagues ravaged the country causing the deaths of large numbers of people. One of the most infamous plagues was the Black Death, which killed off almost one third of the population. In 1665 the Great Plague began in London, killing 68 000 people.

It was not only plagues that ravaged the towns and cities. Fires were also a great problem and in 1666 for example, the Great Fire destroyed a large part of central London. It also destroyed many sub-standard and insanitary dwellings. After the fire a review of the layout of the City of London took place. As a result, new acts were passed to control the rebuilding programme.

During the seventeenth and eighteenth centuries great social changes occurred. The population started to move away from the country into towns. The economy changed from an agricultural to an industrial one. Urban development on a large scale occurred to build dwellings for the factory workers. The local authorities had little control over this development. Industrialists and speculators controlled the use of land, often abusing their power. As a result development was haphazard and uncoordinated.

Parliament became concerned as the conditions worsened. In 1848 the first of many public health acts came into force. The act saw the start of legislative control of public health and town planning.

Today there is a wide range of legislation controlling town planning, public health, factories and building. As our society becomes more complex these acts and regulations are continually revised and their scope expanded.

AN HISTORICAL SUMMARY OF BUILDING LEGISLATION

1189

The first statutory controls relating to buildings in Britain were introduced by Henry Fitz-Alwyn, the first Lord Mayor of London. He produced the first code of byelaws relating to town planning and building construction. Its aim was to reduce the number of fires occurring in the City.

1667

The London Building Act, controlling the rebuilding of the City of London, came into force. It was introduced as a direct result of the Great Fire of London in 1666, its main aim being 'fire protection'.

1774

The Building Act was introduced. One of its provisions was the creation of district surveyors, appointed by city councils to ensure that work was carried out in accordance with the act.

1802

The Health and Morals of Apprentices Act was introduced to reduce, among other things, the working hours of apprentices to 12 hours per day.

1833

A Factories Act was introduced setting up factory inspectors to report on conditions and accidents.

1844

A revised Building Act was introduced. It designated buildings in classes according to their occupation. Fireproofing was enforced for halls, corridors and stairways in public buildings, for the safety of the general public.

1848

The first Public Health Act.

1858

The Local Government Act was introduced to extend the London Building Act to towns and cities throughout Britain.

1875

The second Public Health Act.

1877

The first Model Building Byelaws were produced as a guide for local authorities when producing their own building byelaws.

1933

A Factories Act was introduced to establish the maximum hours that children could work. The act stated that children:

- Up to 9 years of age should not work.
- Up to 13 years of age should not work more than 8 hours per day.
- Up to 18 years of age should not work more than 69 hours per week.

1936

A new Public Health Act was introduced that included the New Model Building Byelaws. They covered the new forms of construction that were being used, e.g. cavity brickwork. Local authorities continued to use them as a basis for their local byelaws.

1948

Town and Country Planning Act.

1961

A new Public Health Act, relevant to the proposed introduction of a set of national Building Regulations, was introduced.

1965

The Building Regulations, a new form of building control, were enacted. It was the first set of regulations covering buildings to apply nationally.

1972

A revised set of Building Regulations was introduced.

1973–5

Amendments to the Building Regulations.

1976

A revised set of Building Regulations was introduced.

1985–91

In 1985 a new set of Building Regulations were introduced, which changed the way in which building construction was defined and controlled in England and Wales. In 1990 and again in 1991 these regulations were revised; the number of sections was also increased, widening the scope of the regulations.

Delegated legislation

Parliament can give to other persons or organisations the power to make legislation. Legislation produced in this way carries the full weight of the law. It is usually in the form of orders, regulations and rules. This type of legislation is called delegated or indirect legislation.

Acts of Parliament now tend to lay down general principles or policy. They leave the production of the often intricate administrative details to other authorities. These authorities are often the ones responsible for putting the acts into effect. For example the Secretary of State for Education may make orders under the Education Act 1944.

FORMS OF DELEGATED LEGISLATION

Orders in Council These are laws enacted by the Privy Council.

Statutory instruments, rules and orders These are drafted by a minister and must usually be approved by Parliament before they come into force.

Byelaws Byelaws require the approval of the appropriate minister before they have any legislative force. They are made by local authorities and national boards that control water, railways, coal etc.

Parliament passes around eighty acts per year. In comparison, the number of statutory instruments produced exceeds 2000 per year.

Factories Act 1961

This act lays down the basic requirements to enable reasonable standards of safety, health and welfare to be achieved. It will eventually be superseded by statutory instruments made under the Health and Safety at Work Act 1974.

The construction regulations are an example of statutory instruments enacted under the Factories Act.

Health and Safety at Work Act 1974

This is the principle act relating to safety, health and welfare on construction sites. Under this act the relevant government minister will enact more regulations as necessary.

A more detailed explanation of its provisions can be found in the chapter on safety.

Construction Regulations 1961, 1966 and 1989

These regulations are produced as an additional set of controls under the Factories Act 1961. They relate specifically to construction operations and cover aspects of safety, health and welfare.

- *Construction (General Provisions) Regulations 1961.* These regulations establish minimum standards of general safety.
- *Construction (Lifting Operations) Regulations 1961.* These regulations cover the construction, maintenance and inspection of lifting appliances used in construction operations.
- *Construction (Health and Welfare) Regulations 1966.* These regulations set down the minimum provisions, for specified numbers of people on site, for first aid, site accommodation, washing facilities and protective clothing.
- *Construction (Working Places) Regulations 1966.* These regulations cover the use, erection and inspection of scaffolding.

- *Construction (Head Protection) Regulations 1989.* These regulations enforce the use of measures to reduce the number of head injuries occurring to people on site.

The Building Regulations 1985–91

The Building Regulations are designed to set minimum standards for all building work and to safeguard public safety and health. They are administered by the local authorities through their building surveyor's department.

Current legislation relevant to construction

The Acts and Regulations listed below are some of the major pieces of legislation currently controlling construction work.

Health and Safety legislation

Abrasive Wheel Regulations 1970
Construction (General Provisions) Regulations 1961
Construction (Head Protection) Regulations 1989
Construction (Health and Welfare) Regulations 1966
Construction (Lifting Operations) Regulations 1961
Construction (Working Places) Regulations 1966
Control of Lead at Work Regulations 1980
Control of Pollution Act 1974
Control of Substances Hazardous to Health (COSHH) Regulations 1988
Electricity at Work Regulations 1989
Explosives Act 1875 and 1923
Factories Act 1961
Fire Certificates (Special Premises) Regulations 1976
Fire Precautions Act 1971
Guard Dogs Act 1975
Health and Safety (First Aid) Regulations 1981
Health and Safety at Work Act 1974
Highly Flammable Liquids and Liquefied Petroleum Gases Regulations 1972
Mines and Quarries Act 1954
Offices, Shops and Railway Premises Act 1963
Protection of Eye Regulations 1974

Reporting of Injuries, Diseases and Dangerous Occurrences Regulations 1985
Safety Signs Regulations 1980
Woodworking Machine Regulations 1974
Work in Compressed Air Special Regulations 1958 and 1960

Legislation controlling construction work

Building Act 1985
Building Regulations 1985
Clean Air Acts 1956,1968
Defective Premises Act 1972
Highways Act 1959
Historic Buildings And Ancient Monuments Act 1953
Housing Act 1988
Public Health Act 1936 onwards complementing the Building Regulations
Town and Country Planning Act 1990
Water Acts 1945 onwards

Employment legislation

Employers Liability (Compulsory Insurance) Act 1969
Employers Liability (Defective Equipment) Act 1969
Employment Protection Acts 1975,1978
Equal Pay Act 1970
Industrial Training Acts 1964,1982,1986
Sex Discrimination Acts 1975,1986
Social Security Act 1975
The Employment Acts
Trade Union and Labour Relations Act 1974
Wages Act 1986

12 The Building and the Environment

THE BUILDING IN THE ENVIRONMENT

Environment

The environment is the natural surroundings of a person, object, building or region.

Built environment

The form of environment constructed by man is called the built environment. It is mainly used to describe buildings, roads and open spaces in built-up areas.

Town

An area of urban development in which people work, live and play.

City

A large area of urban and suburban development. Cities are created by granting a charter to a large town.

Conurbation

Neighbouring towns and villages can expand to such an extent that they merge into one large built-up area. The whole of this built-up area is called a conurbation.

Urban

An area of built environment forming a town or city is called an urban area.

Suburban

The residential areas around the outskirts of a town or city are called suburban areas or the suburbs.

Rural

Areas of land that are used for agricultural purposes or form parts of the countryside are called rural areas.

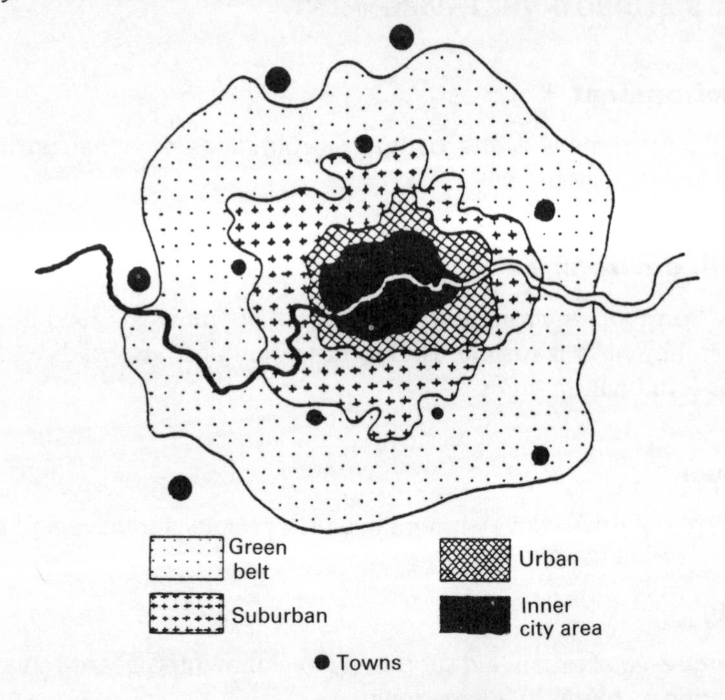

Green belt around a city.

Green belt

The green belt is an area of open land near to and often surrounding an urban area. It is kept open by permanent and severe restrictions on building.

The purpose of the green belt is to:

- Check any further growth of the adjoining built-up areas.
- Prevent neighbouring towns merging into large unplanned conurbations.
- Preserve the special character of our villages, towns and cities.

Derelict land

Disused, spoiled or abandoned land is described as being derelict. The land is often derelict as a result of railway closure, abandoned industrial areas, slum clearances etc. It can give the landscape of the built and natural environment an air of desolation and bleakness. The appearance of derelict land often deters new development and can cause economic depression.

Slum area

Slums are areas of overcrowded, squalid and substandard buildings within the built environment.

Pollution

When unwanted elements contaminate a substance, that substance is said to be polluted. The natural environment is made up of air, water, land, plants and animals. When they become polluted, man's health, food supply and survival are threatened. The natural environment can be contaminated by chemicals, effluent, smoke, fumes, waste products and noise. To try to reduce the amount of pollution, society has many regulations restricting the disposal of contaminants. Often these regulations are insufficient *and* ineffective.

Landscaping

The landscape is the scenery that surrounds us. Often in the built environment the scenery can be drab, depressing, colourless and monotonous. To enliven the scenery in inner city areas attempts are made to improve the landscape.

Changes can be achieved by:

- Clearing, cleaning up and contouring derelict land.
- Cleaning up the face of dirty buildings, especially the stone ones.
- Creating small areas of parkland.
- Planting trees and shrubs.
- Changing the colour and texture of materials such as paving and large expanses of bare brickwork and concrete.

Conservation

Conservation is the management of resources and assets to ensure that they are protected and enhanced. The conservation of existing buildings enables them to continue a useful existence. Conservation involves keeping the appearance of buildings, streets, and open spaces as they are. However, some form of adaptation may be necessary to meet the needs of modern life.

Preservation

When we preserve things we keep them exactly as they are. They are maintained unchanged but saved from decay. A good example of preservation is the application of a chemical to a timber fence post to prevent it rotting away.

Renovation

When something is being renovated it is in the process of being returned to its original state. Historic buildings are often renovated, to restore them to their original condition.

Restoration

The process of repairing something and putting it in a good condition. Old houses, for example, are often restored. In many cases it is cheaper than demolition and constructing a new building.

Town planning

Planning is a physical process that is used to control and harmonise the use of land and the design of buildings.

Town planning enables controls to be placed on the:

Location of industry nationally.
Development of new towns.
Large-Scale control of land use.
Quality of the environment.
Regional economic stability.
Conservation of amenities.
Renewal of inner city areas.

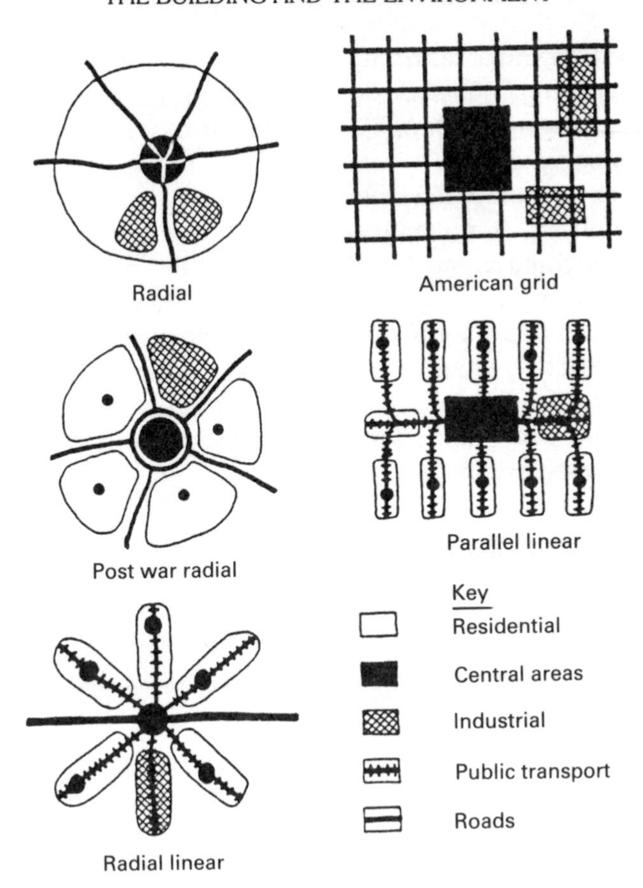

Town plans.

Provision of urban development.
Community development.

Central Government influences town planning via the regional offices of the Department of the Environment and by advice given in the form of Circulars and Guidance Notes. However, the basic responsibility for town planning falls on local government. County councils are responsible for setting the planning framework within which the local authorities produce the detailed policies on which they base their local planning decisions.

The planning system

Local authority town planning departments and the planning committees that control them are involved in a wide range of activities:

(1) Encouraging or restricting the rate and type of expansion e.g. by monitoring population growth and catering for its needs, and by encouraging the development of industrial estates and housing developments.

(2) Regeneration of the older residential areas can be achieved by stimulating the improvement of the houses and their surrounding areas. Local support for this is usually done by designating it a General Improvement Area.

(3) The protection for the future of historic buildings and areas of great beauty are of great importance. To undertake this work, controls are applied to conserve and preserve historic buildings, ancient monuments, sites of special scientific interest and areas of outstanding natural beauty.

Types of planning application

Outline – normally used for major development proposals in order to find out if the proposal is acceptable in principle before more expensive design work is carried out. Approval of this kind usually has a life of three years.

Full – used to obtain full approval to undertake the work required. Approvals of this kind usually have a life of five years.

Change of use – used when an occupier needs to change the use but not the external appearance of a building.

Advertisement – permission is needed to erect advertisements. Those on shop fronts are especially important as they could alter the atmosphere and appearance of a shopping area.

Listed buildings – special approval is needed for work undertaken on listed buildings to ensure that they remain in their original form.

Tree works – if a tree is protected by a Tree Preservation Order approval must be obtained before any work is undertaken on the tree.

Density

The quantity of buildings and the population of areas of the built environment must be controlled. If not there will be overcrowding and overbuilding which will destroy the quality of life and the environment. There are three levels of density – low, medium and high.

LOW DENSITY AREAS

Areas of high rateable value where the buildings are well spaced out and of high quality.
Example Large detached private houses with substantial amounts of land.

Low density development.

MEDIUM DENSITY AREAS

Used for housing developments in towns, or on the outskirts of inner city areas.
Example Detached, semi-detached and terraced housing on well laid out suburban housing estates

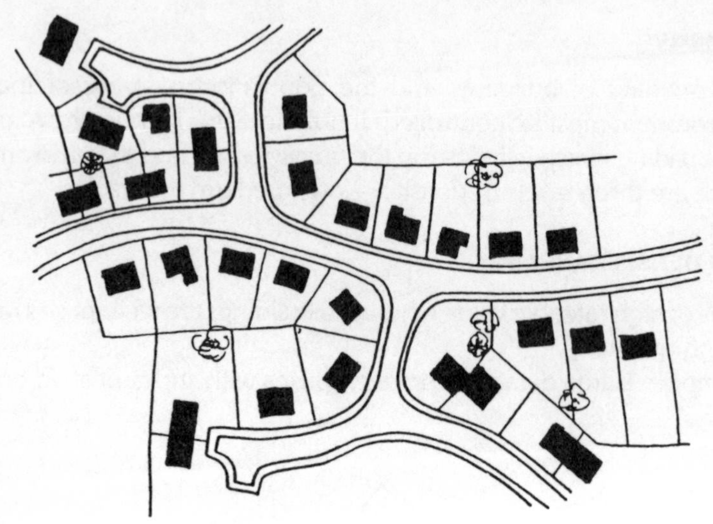

Medium density development.

HIGH DENSITY AREAS

To obtain the most economic use of high value urban land enabling towns and cities to remain compact.
Example town houses and blocks of flats.

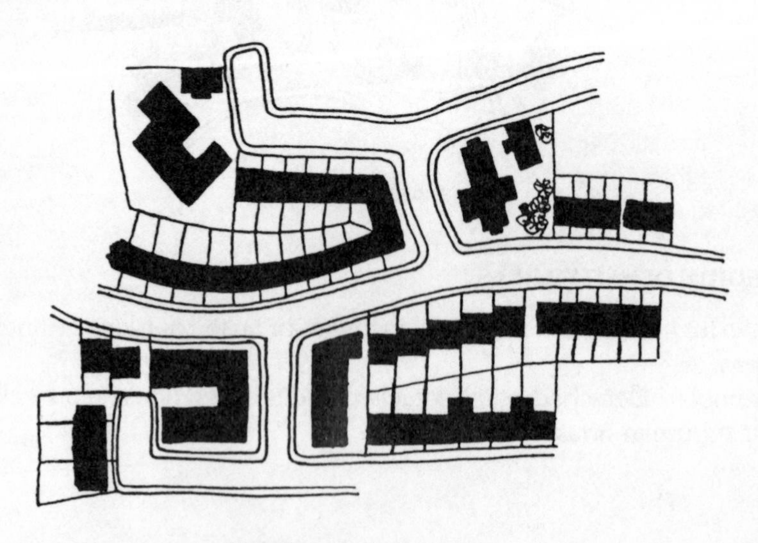

High density development.

Density	Residential dwellings per hectare	Habitable rooms per hectare
low	10 to 15	–
medium	25 to 40	125 to 225
high	50	250

Zoning

Zoning is used to control the use of land by imposing restrictions upon it. Land can be zoned by use, building or population density. In this way, the land can be used effectively and efficiently.

ZONES

(1) Residential.
(2) Business office.
(3) Business shops.
(4) Business wholesale warehouses.
(5) Public buildings, educational and recreational.
(6) Light industrial.
(7) Special industrial.

} Suitable for central areas of the built environment.

13 Case Studies on the Building and the Environment

CASE STUDY NO. 1: THE CAYSTER TOWER

The situation

After a merger with another company, NewStar have become the country's largest banking, finance and insurance group. The new company has two head office buildings in London. Around the country there are a large number of local offices.

An investigation into NewStar's organisation recommends that:

- The company builds a new head office as the present ones are totally inadequate for their needs.
- The local offices be reorganised into regions under the control of a regional office.
- Four regional offices be set up.
- The size of the local offices should be reduced and their numbers increased.
- Local offices should be situated in shopping centres.
- Some of the work done by the local offices should be done at head office.
- The communication of information between head office, regional offices and the local offices should be controlled by a large computer situated at head office.

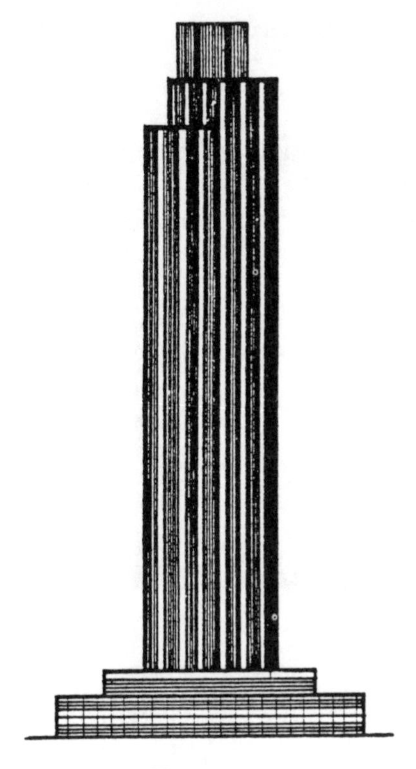

The tower.

After investigating three suitable sites that have been zoned for office development. NewStar have chosen the site in Cayster, a suburb in West London. They propose building a tall tower covered in black solar reflective glass. It will be 200 metres high and have 55 floors containing 30 000 square metres of office space. The company estimate that approximately 2500 people will be employed in the building.

They will be made up of those moving from the existing head offices (1200), those transferring from local offices (400) and new employees (900).

The arguments for the tower

NEWSTAR'S CASE

- The tower will become a landmark on the skyline and a prestigious head office for the company. It will demonstrate publicly that NewStar is a modern, progressive and successful company.

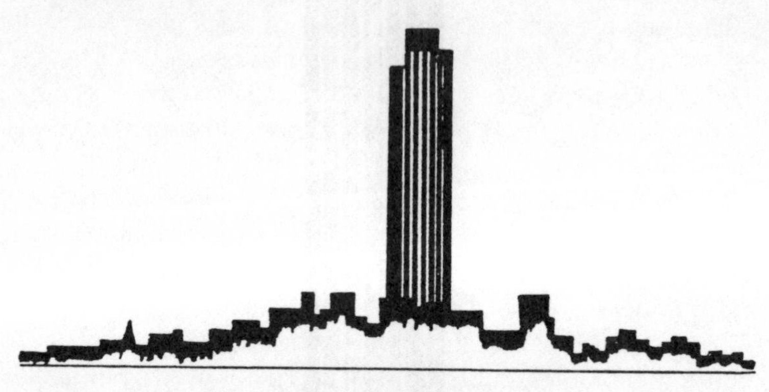

The new skyline.

- The new building will have all the modern facilities available and will be designed to suit the specific needs of the company. By centralising its operations, overheads will become cheaper and internal communications easier.
- Work would be available for 900 new office staff. Some of these jobs would go to local people.

THE LOCAL CASE

- The tower would bring building work into the area and some residents would obtain jobs on the project.
- Permanent new jobs would be created for local people.
- The business prospects for local traders would improve with a possible 2500 new customers coming into the area.
- The 400 employees moving into the area from the regions will require houses. Consequently house prices will rise around Cayster.
- The demand for property will encourage the improvement of the existing housing stock to make it more saleable.
- A large new office block will encourage other firms to move into the area. This will raise office property values and generate more wealth in the area.

The arguments against the tower

- A number of residents and the local paper have formed a pressure group opposed to the development. Their intervention has created arguments amongst the population. Headlines in the paper 'A BLACK TOWER TO PIERCE THE HEART OF CAYSTER' and 'A BLACK DAY FOR CAYSTER' have not helped matters.
- They say the 200 metre high tower of black glass and stainless steel will be an eyesore, visible for miles around.
- It is too large, oppressive and will contain too many people.
- The pleasant atmosphere in this quiet suburban area will be destroyed.
- The tower will encourage more office development in the area.
- Existing road, rail, tube and bus networks will be unable to cope with the strains imposed on them by the large number of commuters.
- Local post offices and shops will not be able to cope with the extra customers. Cafés, public houses, and parks used by the residents will become overcrowded during the lunch hour.
- The protesters argue that the above conditions will destroy the quality of life enjoyed by the residents of the area.

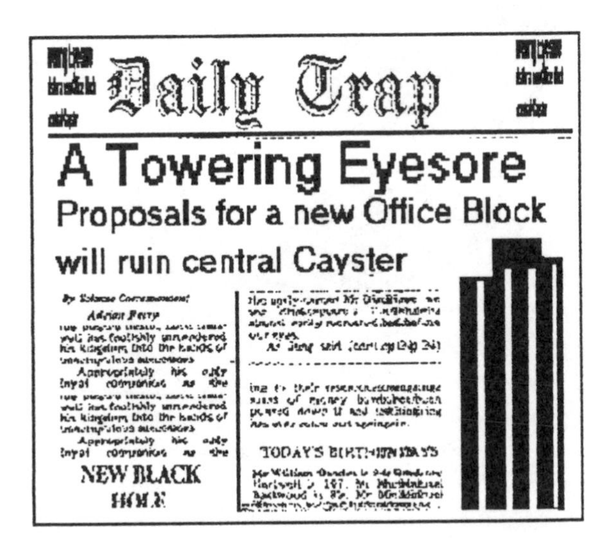

Daily Trap

A Towering Eyesore

Proposals for a new Office Block will ruin central Cayster

NEW BLACK HOLE

The Council's position

- The council is in a difficult position.It would like to approve the planning application for the tower, but is sensitive to local feelings and opinions.
- The council can see a lot of benefits from approving the planning application for the tower. It will receive a large amount of rates from NewStar, money that is badly needed by the council. More office development will generate even more capital. The extra revenues would enable the council to:

  Improve the local road network.
  Re-develop part of the shopping area.
  Renovate a large part of its housing stock.
  Build some new houses.

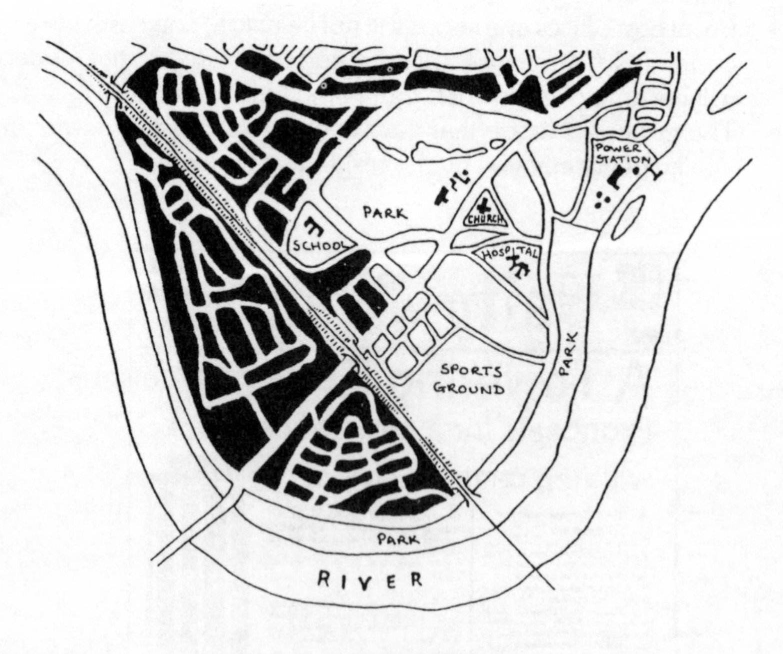

Zone A – Residential areas.

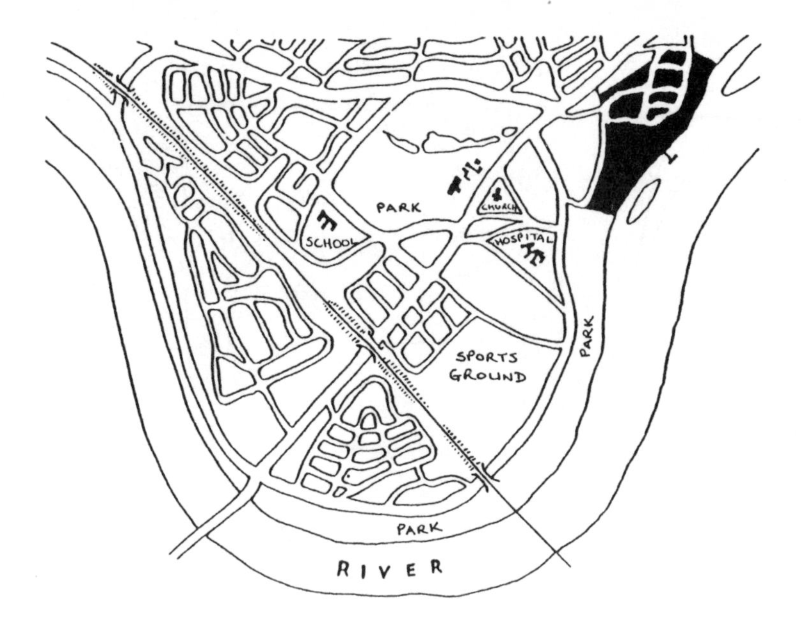

Zone B – Light industrial areas.

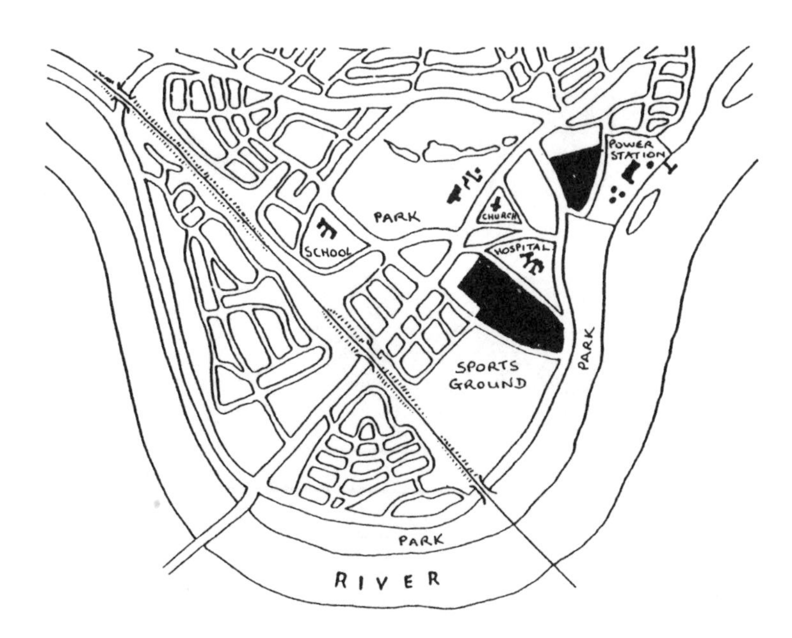

Zone C – Areas of derelict land zoned for development as offices and shopping precinct.

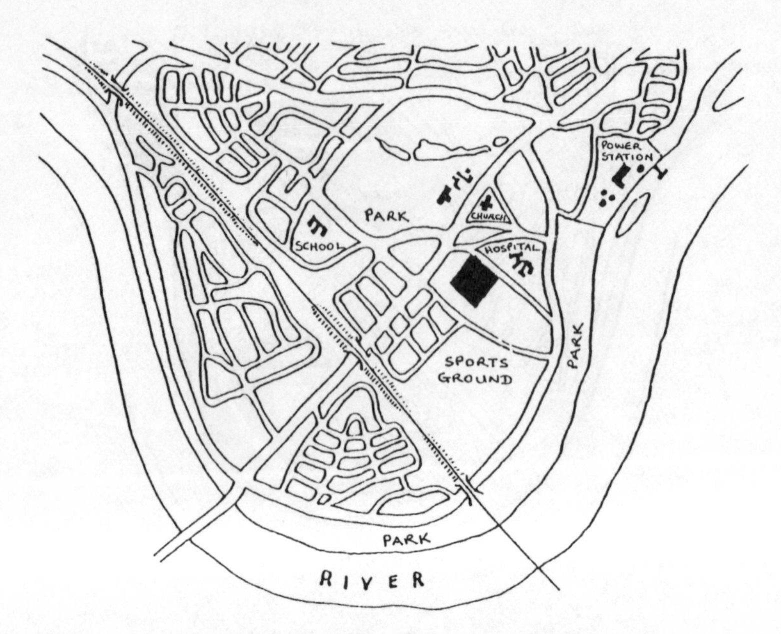

Zone D – Site of proposed office block.

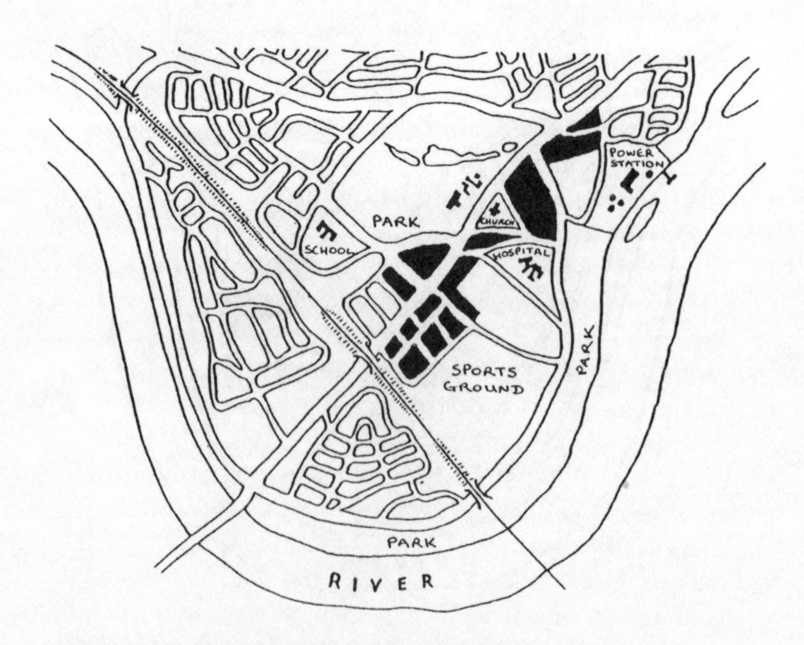

Zone E – Existing office and shopping areas.

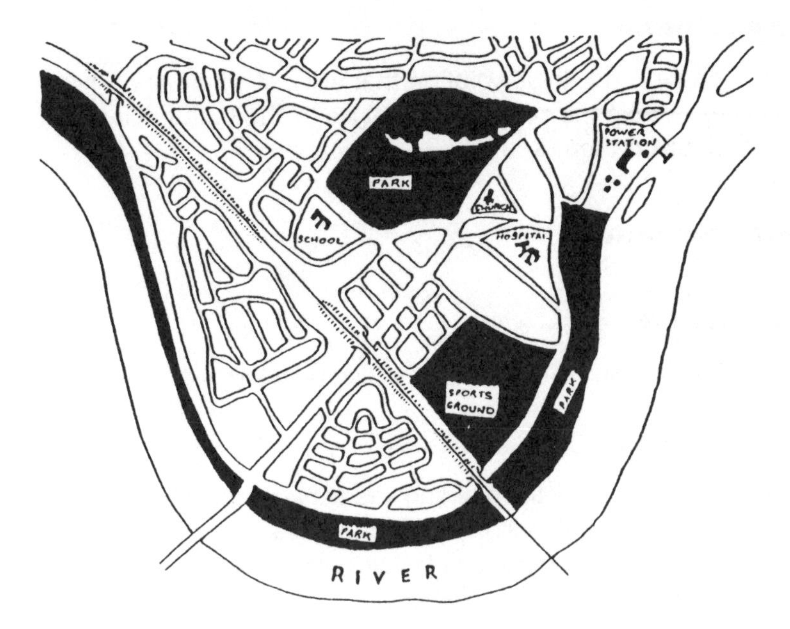

Zone F – Parkland and recreational areas.

Other considerations

- The placing of a large, heavily populated building in the middle of a quiet semi-suburban community, without careful planning, could result in an environmental disaster and damage the quality of life in the area.
- The building may be an architectural masterpiece but will it blend in with the architecture and atmosphere of the area?
- Do more jobs, trade, efficient industry and a revitalised area, have to be obtained at the expense of overloaded road and rail networks, lower living standards for the community and over-stretched postal and telephone services?
- Higher levels of road transport will be necessary. Visitors, commuters and suppliers will need to reach the tower.
- Public transport will come under strain as commuters overload the existing services. Increases in services will raise the traffic flows in the area.
- Existing facilities such as shops, post offices, bus, rail and tube stations may be unable to cope with the extra influx of customers.

The solution

- There is no single solution that will suit all groups. The solution must take the proposed changes in the built environment and consider them as a whole.

The alternatives

ALTERNATIVE 1

- Totally reject the proposed office development.

ALTERNATIVE 2

- Limit the size of the building and persuade NewStar to increase or enlarge its regional offices. Modern telecommunications systems enable communications to take place over large distances with speed, security and reliability.

ALTERNATIVE 3

- Allow the proposed development to take place. Restrict severely any further office development in the area.
- Increase the size and capacity of the public transport systems.
- Use any revenue generated by extra rates from the tower to redevelop the central shopping area, providing better shopping facilities.

ALTERNATIVE 4

- One of the main objections to the tower is the black solar reflective glass cladding. A change of colour to bronze, for instance, could make the tower more acceptable to the community. This extra restriction could be included in Alternative 3 or 4.

COMMENT

NewStar should not be made a scapegoat because it is a large and successful company. Neither should its future development be limited. However, it should not be allowed to destroy a pleasant suburban area and swamp the residents under a mass of new development and large numbers of commuters.

CASE STUDY NO. 2: MRS JONES' RETIREMENT HOME

The situation

Fred Jones owns Coombe Farm, about five miles south of Bridgeford in North Devon. He lives in the 300-year-old historic farmhouse with his wife and four young children.

Grandma Jones lives in a very large old rambling house in the small town of Bridgeford. She is 62 years old and has recently been widowed. After a lot of thought and financial calculation she has come to the conclusion that she can no longer afford to live in her large house on her very small income.

The solution

Grandma Jones has decided that she will buy a small bungalow. The only one that she can find with the facilities that she wants is in a small village about fifteen miles from her son's farm.

When she tells her son of her decision he agrees that it is a good idea to sell her large old house in Bridgeford. However, he does not like the idea of his mother living fifteen miles away in another small village. He has a better idea.

The suggestion

The farmhouse is too small for Farmer Jones and his large family. The children have to share bedrooms and there is not much space for them to play inside the farmhouse. He suggests that:

(1) He builds an extension onto the end of the farmhouse to provide more rooms for his own use and a granny flat for his mother.
(2) To make room for the extension he will have the derelict old barn next to the farmhouse demolished.

The decision

Grandma Jones thinks that this is a very good idea; she will be close to her family and have a small home built to her exact requirements. On hearing her son's solution she asks him to arrange for the work to proceed.

The next step

Farmer Jones instructs his architect to produce a suitable design for the extension. The architect, on hearing Farmer Jones' proposals, expresses doubts as to their success. He does not think that planning permission will be given by the local authority. He suggests that outline planning permission be obtained from the local authority before any major and expensive design work is undertaken.

The refusal

The local authority rejected Farmer Jones' application for outline planning permission. They gave the following reasons:

(1) Coombe Farm is listed as a building of special architectural interest and as such is protected from alteration or demolition. It is a fine example of seventeenth century architecture.
(2) The proposed extension would not match the architectural style of the farmhouse.
(3) The extension would be detrimental to the appearance of the farm as a whole.
(4) Although the barn is not a listed building and protected from demolition, it is of architectural interest and adds to the historical and rural environment of the farm.

The second suggestion

After this setback Farmer Jones comes up with a second proposal. It will not satisfy *his* needs but it will solve his mother's problem. He suggests that his mother builds a small bungalow to suit her exact needs on his farm. He would provide the land free and she could finance the building work from the sale of her house. If there was any money left this could be invested to increase her income.

The second refusal

The local authority once again refused the planning application. They gave the following reasons:

(1) The proposed bungalow would be too close to the farmhouse, which is a listed building.

(2) The proposed bungalow would be built in an orchard on valuable farmland. The reasons given for changing the use of the land were not sufficient to warrant the loss of such a valuable resource as farmland.
(3) The proposed bungalow would spoil the rural environment.

The final solution

In despair Farmer Jones asked his architect to find some other solution. After considering the problem, the architect proposed that Farmer Jones restores his old derelict barn and converts it into a granny flat. The work will cost more than either of the other two schemes, but it may be more acceptable to the local authority.

The final application

The local authority decided that these proposals were acceptable. They approved the application for outline planning permission. The restoration of the old barn, they said, would improve the historical environment of the farm.

The architect's brief

- Produce a plan for renovating the exterior of the barn.
- Restore the interior of the barn to its original condition.
- Design an interior layout for the barn containing a small two-bedroom flat suitable for a senior citizen and a large playroom.

INTERIOR

The interior design must be:
Simple.
Functional.
Easy to keep clean.
Cheap to run.
Easy to maintain.
Erected quickly and easily.
Employing construction techniques that do not destroy the appearance of the barn.

ROOMS

The rooms must be light and airy, making full use of the good views of the surrounding countryside from the site. They would consist of:

Hall.
Lounge/diner.
Kitchenette.
Bathroom.
Toilet.
Double bedroom.
Small bedroom.
Large playroom.

FACILITIES

A number of modern facilities should be included in the design to ensure that the dwelling provides a high standard of comfort, such as:

Central heating.
Double glazing.
Fully fitted kitchenette.
Fitted wardrobes in the bedrooms.
Non-slip floors in the bathroom, toilet and kitchenette.

CASE STUDY NO. 3: A BETTER PLACE TO LIVE

The situation

Dufford is one of a number of small commuter towns within easy reach of the city. It is often the target for local humour because of its poor reputation. The town seems to consist of street after street of dirty stone-fronted houses. Visitors to the town say that it is dirty, dismal and has an air of decay about it.

House prices are low compared to those in nearby towns. Even so, the availability of cheap housing has failed to tempt people to move into the area. Many houses are up for sale, often standing empty for long periods before they are sold. The lack of movement in house sales is probably due to the poor condition of the houses. They are generally in a bad state of repair. The house owners seem to have little or no interest in improving the condition of their property. As a result, the town's only

DIY shop has closed down through lack of business. It seems as if the town is slowly decaying through lack of interest and care.

Neighbouring Carrbridge is a town of comparable size and has many residential areas similar in age and architecture to those in Dufford. However, it does not seem to suffer from the problems that occur in Dufford. It is a desirable place to live, with a buoyant housing market. Visitors to the town often remark on how the town seems to be bright, cheerful and alive.

The decision

The local authority in Dufford is concerned about the poor reputation of the town. Anxious to find ways of improving the image of Dufford they commission an investigation. The job of the investigators is to find out why the neighbouring towns are more popular.

The report

SUMMARY OF INVESTIGATIONS IN CARRBRIDGE

The local authority in Carrbridge has produced a number of schemes designed to improve the town. These schemes are designed to improve the built environment in Carrbridge. The most effective schemes cover the following areas:

Clean air.
Landscaping.
Home improvements.
Local industry.

CLEAN AIR

By introducing smokeless zones the level of atmospheric pollution has been reduced. Consequently, the buildings in the town remain cleaner longer. Cleaning the dirty stone buildings gives them a new bright appearance. Because the air is cleaner there are fewer corrosive elements in the atmosphere to damage the buildings.

LANDSCAPING

Carrbridge used to have many areas of derelict land caused by the closure of local industries. To improve the appearance of the urban scenery a landscaping programme was put into effect. The derelict land was cleared of rubbish, levelled and then landscaped. The total extent of the landscaping only involved grassing the land and planting a few trees.

The scheme was popular and effective in improving the urban landscape. The success of the initial scheme encouraged the local authority to expand into the following areas:

- Landscaping wherever possible all derelict land.
- Tree planting in suburban streets and urban civic areas.
- The creation of mini-parks and gardens on small areas of unused land.

HOME IMPROVEMENTS

To encourage home owners to improve the condition of their property, the local authority set up a scheme to give home improvement grants. These were available for:

- Re-roofing the old slate roofs with tiles.
- Loft insulating.
- Putting in new bathrooms and toilets.
- Installing central heating or smokeless fires.
- Cleaning the front elevations of dirty stone buildings.

The local authority also started schemes to improve the condition of its council houses, bringing them up to present-day standards.

LOCAL INDUSTRY

To attract new industry into an area, it is necessary to have an ample stock of suitable factory space. To provide this factory space the local authority has built several small industrial estates. Artificially low rents are charged to encourage firms to move into the area and help new small businesses get off the ground. As a result of these policies the amount of industry and job prospects in Carrbridge have improved.

Recommendation

The investigators stated in their report that the condition of the built environment is the responsibility of the community. Its improvement can be expensive but the benefits to the community are enormous.

The report recommended that Dufford local authority set up schemes to improve the built environment and to encourage the local community to take an active part in improving their environment.

14 History of Building

Man has had to construct some form of dwelling since the birth of civilisation. At the very beginning he could only construct very simple shelters to protect himself from the elements and wild animals. As his standard of living and civilisation improved his dwellings also improved.

The construction of buildings has become a feature of each period of our history. The different architectural styles are a measure of the way people lived, their advances, disasters and social standards.

The design and construction of buildings show the effects and influences on man of:

Climate.
Materials.
Social pressures.
Advances in technology.
Religion.
Work.
Pleasure.

THE ROMANS

During the Roman occupation of Britain many changes occurred, the most noticeable in building construction. Before the invasion the majority of the population lived in round wattle huts. Roman technology brought the following innovations:

Straight paved roads.
Stone buildings.
Public baths.
Temples.
Defensive walls.
Brickwork.
Tiled roofs.
Mosaic.

Paved floors.
Square rooms.
Plumbing.
Baths.
Underfloor central heating.
Water viaducts.
Lead was mined and tiles and bricks were manufactured in large quantities.

After the collapse of the Roman Empire the manufacturing and construction skills introduced by the Romans were lost, and the local population reverted to their native traditions.

PERIOD: SAXON
DATE 650–1066

The events

597 St Augustine lands in England to spread Christianity.
883 Winchester is capital of England.
1066 Norman invasion.

The people

KINGS

Alfred the Great built the first English navy.
Edward the Confessor.
Harold was beaten at the Battle of Hastings.

St Augustine. } Missionaries.
St Columba. }
The Venerable Bede – a famous monk, writer and historian.

The buildings

Early Saxon buildings were made from sloping timbers joined together at the top by a ridge. They were thatched with straw, heather or reed. The open ends were blocked off with wattle panels. These would be removed during the day to allow ventilation and access. After the Viking

A Saxon house made from timbers covered with reed thatch.

invasions the cruck frame was introduced, which gave more headroom inside the dwelling. This development reduced the need to dig out a pit in the floor of the dwelling to obtain sufficient headroom.

The Saxons only constructed a few stone buildings, which were mostly churches, abbeys, cathedrals and towers.

Standard construction techniques

Cruck-framed buildings.
Wattle and daub infill panels.
Thatched roofs.

Innovations

Lead was used occasionally for covering roofs.
Some churches had glazed windows.
At the latter end of this period some stone buildings were constructed.

The materials used

Rough hewn timber.
Heather, straw or reed for thatching.
In 675 a monk called Bede recorded the use of glass to glaze a church at Monkwearmouth. It was installed by French craftsmen.
By 900 the use of stained glass was common.

PERIOD: NORMAN

DATE 1066–1189

The events

1066 Norman invasion.
The Battle of Hastings.
1069 The Feudal system introduced.
1077 London burnt down.
1086 The Doomsday Book.
1091 Start of the Crusades.
1161 London, Canterbury, Exeter and Winchester were burnt down.
1166 Start of common law and the jury system.
1189 First London building regulations introduced by Henry Fitz-Alwyn, first Lord Mayor of London.

The people

KINGS

William 'the Conqueror' of Normandy.
Henry II.
Richard I 'the Lionheart'.

Thomas à Becket, Archbishop of Canterbury, assassinated 1170.

The Norman manor house built of stone at Boothby Pagnall, near Grantham.

The buildings

The Normans built many castles in England and from them enforced their rule over the surrounding countryside. On their smaller estates and farms they built fortified manor houses to protect themselves and their livestock.

The manor houses had two floors. The ground floor was used for storage and animals. The upper floor was one large room which everyone shared. It was reached by an outside staircase. The cooking and heating was by a central open fire whose smoke escaped through a hole in the roof. After 1150 the castles and manor houses where built mostly of stone.

Norman architecture is easily recognised by its massive appearance, large round columns and the heavy semi-circular arches spanning them.

Standard construction techniques

Narrow window openings.
Stone walls infilled with rubble.
Semi-circular arch.
Barrel-vaulted roofs covered with an outer timber roof.

Innovations

Castles.
Fortified manor houses.

The materials used

Timber.
Stone which was often imported to satisfy demand.
Glass in the windows of churches and important houses.

PERIOD: GOTHIC

EARLY ENGLISH, DATE 1189–1307
DECORATED, DATE 1307–1377
PERPENDICULAR, DATE 1377–1485

The events

1212 The use of thatched roofs on houses in London banned to reduce the risk of fire.
1215 Magna Carta.
1265 Simon de Montforts' Parliament.
1245–70 Westminster Abbey built.
1349–52 Black Death killed almost one third of the population.
1381 The Peasant's Revolt led by Wat Tyler.
1415 The Battle of Agincourt.
1476 The first printing press in England was used by William Caxton.

The people

KINGS

John.
Henry III, IV, V and VI.
Robert the Bruce, King of Scotland.
Richard II, III.
Edward III, IV.

Geoffrey Chaucer, one of the earliest English poets.
William Caxton the printer.

The buildings

Dwellings, although still simple structures, became more permanent with the improvement of the cruck frame. Greater headroom and a larger floor area were possible with the introduction of side walls. Windows were still mostly of timber slats.

A cruck frame structure.

A cruck frame cottage.

Manor houses developed from a fortified hall with a single room shared by everyone, into larger, more complex, buildings. More rooms were added to provide kitchens, stores, chapels, stables and accommodation for the servants. The large hall remained the centre of the house, and it was the place where everyone met and ate. The lord of the manor and his family sat apart from the rest of the occupants, on a raised platform at one end of the hall.

The Gothic period is noted mostly for the development of the pointed equilateral Gothic arch. The arch was first used at the intersections of barrel vaults. Later it was used in roof construction, and in windows, from about 1200.

This period of architecture is noted for the style of its arches and windows.

EARLY ENGLISH

Improved stone-cutting skills and a greater understanding of the principle of forces allowed the builder to construct taller, slender arches. At first windows were tall thin lancets. Later they contained groups of lancets under a main arch. The space at the top of the arch was pierced with tracery.

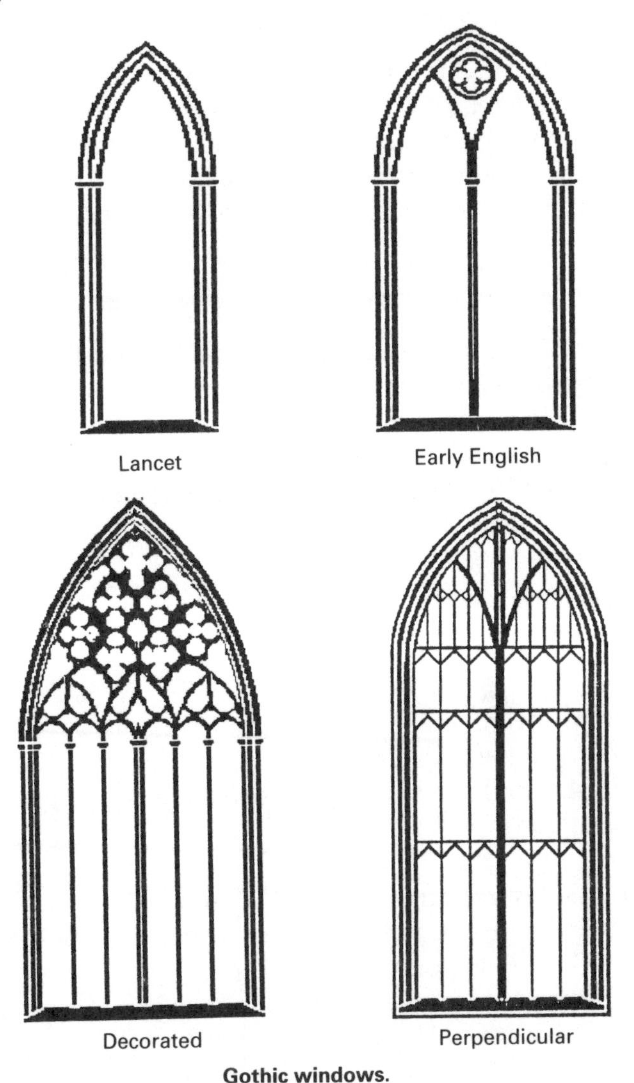

Gothic windows.

DECORATED

In the fourteenth century major buildings became more elaborate, with large amounts of decoration. Arches became broader and less pointed. Larger windows were constructed using mullions as vertical divisions. Elaborate tracery based on geometrical shapes, mouldings, carved foliage and ball flowers decorated the inside of buildings. On the outside of the buildings heavily moulded and decorated arched niches were constructed to contain statues.

PERPENDICULAR

The perpendicular style consisted of uninterrupted parallel lines running vertically from floor to roof. These vertical lines splayed out into great curves of fan vaulting in the roof. The arch became lower and flatter. Other types of buildings were now being built, such as guildhalls, schools, hospitals, and colleges at Oxford and Cambridge.

Construction techniques and innovations

Stone was cut more accurately with the introduction of the chisel, which replaced the axe.
Pointed Gothic arch.
Ogee arch.
Buttresses and flying buttresses were used to strengthen the structure.
Decorative carved stone.

The materials used

Stone.
Glass.
Lead roof covering.
Tiles hung with wooden pegs.
Pigs' bristles and squirrels' tails used in the manufacture of brushes.
Plaster of Paris.

1328 the first record of slates being used.
1325 home produced bricks. Prior to this date most bricks were Flemish, imported from Holland.
1477 Regulations standardised the size of tiles at the equivalent of 265 x 160 x 15 mm.

Cost of labour

First reference to the crafts of the joiner and the bricklayer.
Pay for a master carpenter was 3 pence a day.

Cost of materials

1239 Linseed oil for paint 2 shillings a gallon.
1286 Glass cost 6 shillings for 2.5 pounds (approx. 1 square foot).
1290 Paint pigments: blue 1 shilling and 7 pence an ounce, green and red ochre 1 shilling a pound. Gold leaf cost 5 shillings for 100 leaves.
1350 Wooden shingles cost 5 shillings a thousand.
1375 The slates used on Rockingham Castle cost 8 shillings a thousand.

PERIOD: TUDOR
DATE 1485–1558

The events

1534 Henry VIII breaks with the Pope in Rome and becomes the head of the Church of England.
1536 Dissolution of the monasteries.

The people

King Henry VIII.

Sir Thomas More.
Cardinal Wolsey.

The buildings

The buildings of this period were of timber-framed construction, the framework being infilled with brick or whitewashed plaster panels. Upper floors projected out over the lower ones. The projecting part of the structure is known as a jetty. It was commonly used on buildings in towns, where land was scarce, to gain extra floor space.

A Wealden style timber framed house.

During this period a style of house called the Wealden house was common. It was so called because of its popularity in the Kent and Sussex Weald.

The house was of timber-framed construction. A passage entry gave access to a large open hall in the middle of the house and a kitchen. At the other end of the hall a flight of stairs gave access to a private upper room called a solar. Heating was still by open fire in the middle of the hall. The roof would be tiled, or on poorer dwellings, thatched. This type of house was usually built by wealthy merchants and craftsmen in the towns, or by yeoman farmers in the countryside.

Construction techniques and innovations

Timber-framed houses with jetties at upper floor levels.
Small windows.
Oriel and bay windows.
Bricks laid in a variety of decorative patterns.
Chimneys in a variety of twisted forms.
Door locks and padlocks.
Lead piping for water supply.
Timber panelling with linenfold pattern.

The materials used

Bricks in large quantities.
Lead piping.

PERIOD: ELIZABETHAN
DATE 1588–1603

PERIOD: JACOBEAN
DATE 1603–1620

The events

1588	Spanish Armada defeated.
1600	East India Company formed.
1603	King James VI of Scotland becomes King James I of England and Scotland, the two countries being united under one King for the first time.
1605	The Gunpowder Plot.
1620	The Pilgrim Fathers set sail for America in the Mayflower.

The people

Queen Elizabeth I.

Sir Walter Raleigh.
Sir Frances Drake.
Guy Fawkes, leader of the Gunpowder Plot.

The buildings

As the architectural style of this period developed, the buildings became more decorative. By using extra shaped timbers in the framework elaborate patterns were obtained.

An Elizabethan style timber framed house jettied out at first floor level.

A Jacobean brick – built house, rectangular in plan. The walls are broken only by the horizontal courses at each floor level.

Up to Tudor times most of the major buildings and large houses were build by nobles and the clergy. During the reign of Queen Elizabeth I the country became more prosperous. Towns began to grow in size. As the merchants and traders became wealthier they started to build more substantial houses and shops. The very wealthy built large elaborate mansions.

During the reign of King James the Dutch influence in English architecture gave way to a more classical style.

Construction techniques and innovations

Larger windows with square mullions and transoms.
Timber window frames with diamond-shaped leaded lights.
French windows.
Tile hanging on walls.
Steep roof surfaces with gables and turrets.

The materials used

Brick.
Stuccoed plaster.
Oak panelling.

Cloth and paper wall coverings.
Tiles.
Slates for roofing.
Thatch used only on very poorest of dwellings.

PERIOD: BAROQUE
DATE: 1620–1700

The events

1642	The Civil War.
1649	Charles I executed.
1653	Oliver Cromwell became Lord Protector of England.
1660	Restoration of the monarchy.
1662	Hearth tax 2 shillings per fireplace.
1665	Great Plague of London.
1666	Great Fire of London.
1667	Act for the rebuilding of the City of London.
1690	The Bank of England was established.
1695	Window tax.

The people

Inigo Jones Sir Christopher Wren	architects.
Grinling Gibbons	woodcarver.
Tijou	wrought-iron smith.

The buildings

The Baroque period was a time of re-birth in architectural design. There There was a renewed interest in the classical forms of Greek and Italian architecture. Architects often visited Italy to study the classical buildings. Two of the most prominent architects of this period were Inigo Jones and Sir Christopher Wren.

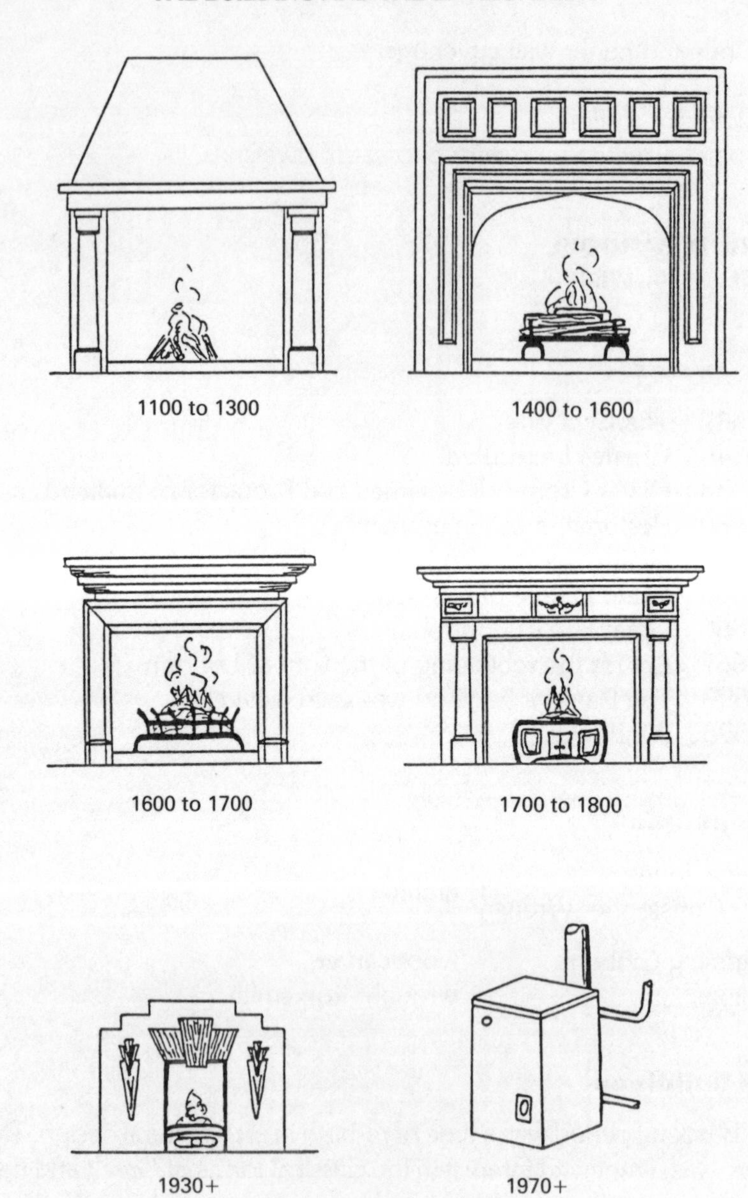

Fireplaces.

Mompesson House, Salisbury, an example of the Queen Anne style of architecture. The front elevation is symmetrical with all its features carefully proportioned. The only decoration is the carved stonework over the front door.

Buildings were constructed on a rectangular plan with symmetrical elevations. Dormer windows were used to give light to the rooms in the steeply pitched hipped roofs.

An example of a house design common in the second half of the seventeenth century. The roof is steeply pitched with dormer windows. The eaves are decorated with carved and painted wooden cornices.

Construction techniques

Hips instead of gables.
Bow windows.
Slate roofs

Innovations

Dormer windows.
Heavily decorated and painted plasterwork.

1676 Vertical sliding sash windows used on Chatsworth House in Derbyshire.

The materials used

Wrought iron.
Cast plate glass.
Cast lead.
Milled lead

PERIOD: GEORGIAN
DATE 1720–1800

PERIOD: REGENCY
DATE 1800–1840

The events

1712 Wallpaper tax.
1754 Furniture design book published by Chippendale.
1711 The first canal was built.
1769 James Watt's steam engine.
1775 American War of Independence.
1779 The world's first iron bridge was built at Coalbrookdale in Shropshire.
1784 Brick tax.
1789 Start of the Industrial Revolution.
1800 Population started to move into the towns from the countryside.

1801 First census in Britain.
1807 Slavery abolished.
1816 First London street lit by gas.
1821 Faraday invented the electric motor.

The people

KINGS

George I, II, III and IV.

John Wood, the architect who designed Bath.
Robert Adam, architect.
Capability Brown, landscaper gardener.
James Brindley, Thomas Telford } civil engineers.
Thomas Chippendale, George Hepplewhite, Thomas Sheraton } furniture makers .

The buildings

The increase in trade and improved methods of farming created a great deal of new wealth. As in the Elizabethan period, a lot of this wealth was spent on building large town and country houses. The design of the houses tried to imitate the classical temples of Italy. Many of the great country houses, like Chatsworth, were built during this period.

By the middle of the eighteenth century the buildings had become very ornate and elaborate. Rooms were large and very ornate with lofty ceilings.

The architect Robert Adam imposed order and style into the period with his designs. His elegant houses had plain simple exteriors with severe lines. The interiors were heavily decorated with ornate and intricate patterns.

The once popular formal gardens gave way to more natural landscapes created by landscape gardeners. The most famous of these was Capability Brown.

Skilled workers still lived in small traditional cottages with inadequate sanitation. The poor lived in squalid conditions in primitive hovels.

A Georgian house of light coloured stucco and dark brickwork, built by the Adam Brothers in Adam Street, London.

Workers cottages.

Construction techniques

Stucco work, brickwork rendered and painted to look like stone.
Vertical sliding sash windows.

Innovations

The town house.
Lavish interior decoration.
Canals.
Woodscrews.
Steam heating invented in 1784 by James Watt.

The materials used

Materials became cheaper and more readily available. This was due to the improvements, made by the use of canals, in transporting heavy materials such as slate, stone, bricks etc.
Imported Chinese wallpapers.
Mahogany and walnut, which arrived in the country as ships' ballast, was used for the manufacture of furniture.
Widespread use of slates for roofing.

1796 Cement.
1800 Britain became one of the largest producers of copper.

PERIOD: VICTORIAN
DATE 1840–1890

The events

1825 Stephenson's Rocket, first steam railway from Stockton to Darlington.
1830 Manchester to Liverpool railway.
1850 Brick tax removed.
1851 The Crystal Palace was built from steel, cast iron and glass.
1868 Elliptical steel arches spanning 72 metres were erected at St Pancras railway station.
1889 The Eiffel Tower in Paris was built of steel.

During the Industrial Revolution the population continued to move to the towns and cities from the countryside.

The people

Queen Victoria

Disraeli
Gladstone } prime ministers.
Peel

William Archer
Augustus Pugin
Norman Shaw } architects.
Gilbert Scott
Phillip Webb

Charles Darwin
Michael Faraday } scientists.
Joseph Lister

George Stephenson
Mark Isambard Brunel } engineers.

Thomas Cubitt, building contractor.

The buildings

The Houses of Parliament.
The Natural History Museum.
The Foreign Office.
St Pancras Station.
Crystal Palace.
The Royal Albert Hall.

Great Britain became a powerful and wealthy nation, partly due to the Industrial Revolution. Consequently, more and more buildings were erected. Some of these buildings reflected the power and wealth of the country, others the inequalities and deprivations of its society.

The Victorian style of architecture imitated both the classical and Gothic periods. Most of the domestic buildings had some form of Gothic influence in their design, usually windows with the Gothic arch and lots of ornamentation.

The buildings were constructed from a variety of materials. They ranged from stone, red and terracotta coloured bricks, to white, green and multicoloured glazed tiles.

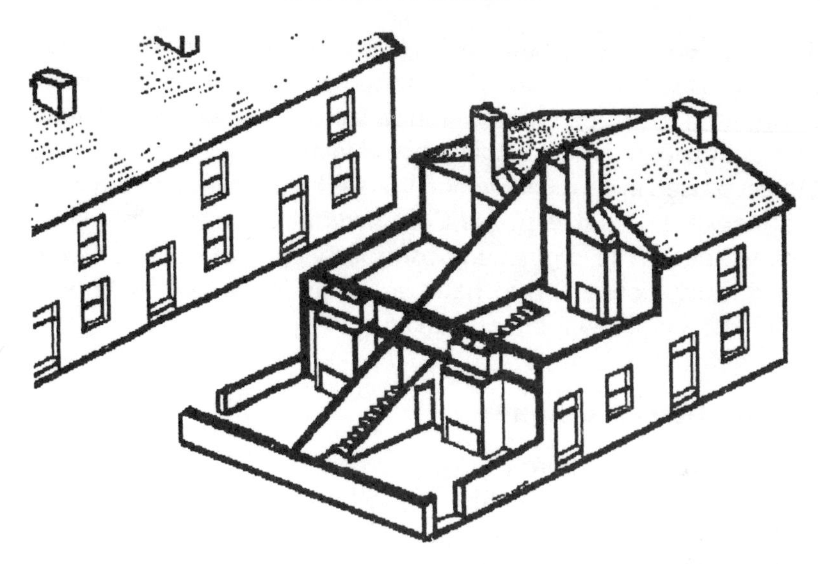

A typical layout of a Victorian back-to-back terrace house. Each house has two or three rooms, one room on each floor.

The movement of people into the industrial areas from the countryside continued. As a result, more and more houses had to be constructed to accommodate them. The houses built by the industrialists for their workers were cheap and shoddy. They contained the minimum possible amount of accommodation and had few facilities.

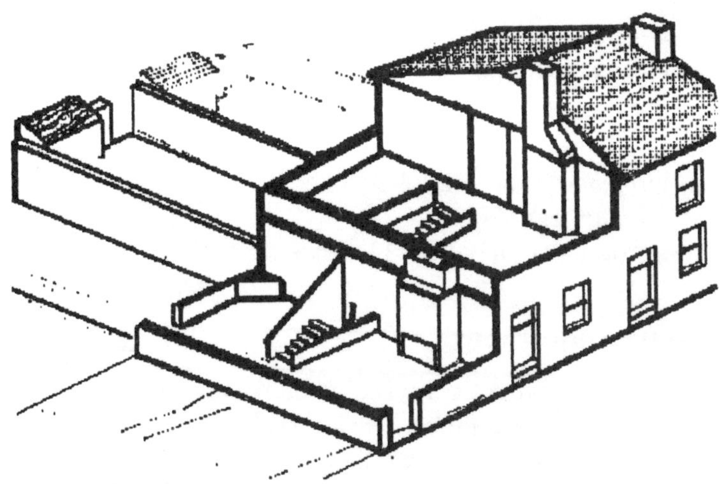

A typical terrace of two storey, late Victorian workers houses. The houses have four rooms, two up and two down. The lavatory was usually outside, often at the bottom of the back yard.

Constructed row upon row in terraces they soon turned into squalid slums. Many of the two- and three-roomed houses were built back to back in terraces. They suffered from a lack of ventilation and light. A group of houses shared a single tap and lavatory.

An act passed in 1875 prevented houses being crammed together. It gave the local authorities power to enforce minimum requirements on house builders. The regulations controlled the standard of workers' housing. Each house had to have, as a minimum, two bedrooms, a lounge, kitchen, its own tap and lavatory.

Construction techniques

Structural use of cast iron.
Less hand work and more use of machines.
High standard of craftsmanship.
Use of high-quality materials.

Innovations and materials

1810	Cast iron drain pipes.
1824	Portland cement.
1830	Reinforced concrete.
1833	Structural use of cast iron.
1840	Concrete tiles.
1845	The oil lamp.
1850	Rolled steel joists
	Cylinder printing of wallpaper.
1855	Gas mantle.
1860	Linoleum.
1870	Fireplaces built with backboiler.
	Greater use of glass.
	Introduction of bathrooms, lavatories and hot water systems.

The building contractor

As buildings became more complicated there was a need for someone to coordinate the various trades. The first building contractors started to operate in this period. Most of their work was speculative house building. The man reputed to be the first large-scale building contractor was Thomas Cubitt, who lived from 1788 to 1856. His building firm is still in existence today.

PERIOD: TWENTIETH CENTURY

The events

1914–18 First World War.
1919 Housing Act enabling councils to erect council houses for rent.
1921 Day release classes started to enable people at work to go to college.
1926 General Strike.
1939–45 Second World War.

The people

Edwin Lutyens
William Morris
Charles Vosey
Sir Basil Spence
Colonel R. Seifert
} architects.

The buildings

Liverpool Cathedral.
Centrepoint.
Post Office Tower.
Natwest Tower.
Humber Bridge.

PERIOD 1890–1914

The Victorians constructed highly ornate buildings with excessive amounts of decoration. At the turn of the century there was a move towards the more simple styles of architecture. There was also a reaction against the harshness and poor quality of Victorian housing in the congested cities.

New suburbs were constructed on the outskirts of the urban areas. Some of these were called garden suburbs. A garden suburb was an area of low or medium density development that had a large amount of green spaces and trees. Each semi-detached house was situated in a wide tree-lined street, and had its own front and back gardens.

PERIOD 1914–1939

Up to the end of the First World War a great deal of care and skill was lavished on the homes of the wealthy. Little care was taken over the homes of the majority of the population.

In the period between the two world wars changes started to occur in the design and construction of buildings. The standard and quantity of houses improved, reflecting the economic and social changes taking place in the country. Architectural styles imitated those of the past. Houses built in the mock Tudor and mock Georgian styles were commonplace.

Four million new houses were built at a rate of 334 000 new houses per year, between 1919 and 1939. By 1939 approximately one third of all houses were newly built. The typical dwelling was a brick-built semi-detached house with three bedrooms, lounge, dining room, kitchen, and inside toilet and bathroom.

Many of the new housing schemes took the form of ribbon development along the sides of railway lines and arterial roads. Development of this kind created an uncontrolled erosion of the countryside around the urban areas. One of the results of this kind of unchecked growth was the strengthening of the town and country planning acts and the creation of the green belt.

New methods of construction and materials were used to improve building standards. Most new buildings were of cavity wall construction, often with a blockwork inner leaf. Many large buildings were of steel or concrete frame construction. New techniques and materials were used to improve the standard of sanitation and drainage.

PERIOD 1939 to present day

After the Second World War many towns and cities had large areas devastated by bomb damage. Rebuilding programmes took place and many buildings in the 'modern style' were erected. The architectural style during this period is best described as a geometrically simple steel- or concrete-framed box. The buildings have large windows giving the interiors a light and airy atmosphere. The exteriors are very plain with no ornament or decoration. Examples of this style of architecture can be seen in the many multi-storey office blocks in our city centres.

Large numbers of new houses have been built since 1945. They were needed to replace those destroyed during the war, and the squalid substandard Victorian slums, and also to provide homes to cater for the rapid growth in the population. These new houses were built on large suburban housing estates or in new towns such as Basildon, Milton Keynes and Cumbernauld.

One of the solutions to the housing problems of the inner city areas was the erection of tall blocks of flats. These tower blocks soon proved to be social and economic disasters. Many of them have been rendered unfit for occupation and demolished, some because of design and construction defects, others because of the social and vandalism problems they caused.

The buildings described in chapter one are examples of the architectural style of this period.

Construction techniques

Steel- and concrete-framed buildings.
Precast concrete construction.
Prefabrication of units and sections of buildings.

Innovations

1900 British Standard for Portland cement.
1904 First reinforced concrete building.
1909 Bakelite.
1914 Stainless steel.
1933 Polythene.
1938 Nylon.

Garden cities.
New towns.
Prestressed concrete.
Lightweight concrete.
Cavity wall construction.
Large picture windows.
Casement windows.
Double glazing.
Gas lights.
Central heating.

Bathrooms in every house.
Mechanical plant.
Prefabrication.
New powerful adhesives.

The materials used

Laminated timber.
Plywood.
Chipboard.
Plastic sheets.
Alloys.
Aluminium.
Plastic pipes and fittings.

Many of these new materials are fully described in chapter 4 on materials.

Industrial Studies Syllabus

City and Guilds of London Institute

Elements, functions and principles of construction

Recognise the different forms of buildings and the main principles involved in their construction.

Recognise the main elements of a building (e.g. substructure, superstructure, cladding, partitioning. finishing, weathering, services).

Demonstrate a knowledge of the function and principles of each element.

Demonstrate a knowledge of the main groups of building materials and their uses.

Sketch given elements and show the relative position of components.

Recognise common faults, defects and failures in buildings.

The process of building and the building team

Demonstrate a knowledge of the stages in designing a building, from defining function to production of working drawings.

State the role of each of the main members of the design and construction team (including the client).

Indicate the main requirements for establishing and maintaining good working relationships within the construction team.

Demonstrate a knowledge of the structure of contracting firms of various sizes and of the roles and responsibilities of employers and employees.

Demonstrate a knowledge of the main principles of organisation and management of jobs.

State the main stages in the construction of buildings.

Outline the general requirements for safety and welfare in construction work.

The Building and the Environment

Understand the need for conservation and improvement of the environment.

Indicate how buildings and construction may enhance or mar the environment.

Show how the need for building changes with economic and social change.

Recognise and name the main types of historical buildings and relate their design to the social needs at the time of construction.

Demonstrate a knowledge of simple aesthetic principles relating to the interior and exterior appearances of buildings.

Demonstrate a knowledge of the social effects of building obsolescence and methods of renewal and renovation.

Demonstrate a knowledge of the ways in which society controls building work through planning and public health law.

Demonstrate a knowledge of the purpose and aims of training and further education and of their value to the individual, the industry and the community at large.

Index